Why was it that Francis Bacon, trained for high political office, devoted himself to proposing a celebrated and sweeping reform of the natural sciences? Julian Martin's investigative study looks at Bacon's family context, his employment in Queen Elizabeth's security service and his radical critique of the relationship between the common law and the monarchy, to find the key to this important question. Deeply conservative and elitist in his political views, Bacon adapted Tudor strategies of state management and bureaucracy, the social anxieties and prejudices of the late-Elizabethan governing elite, and a principal intellectual resource of the English governing classes – the common law – into a novel vision and method for the sciences. Bacon's axiom that 'knowledge is power' takes on far-reaching implications in Martin's challenging argument that the reform of natural philosophy was a central part of an audacious plan to strengthen the powers of the Crown in the state.

Francis Bacon, the State, and the Reform of Natural Philosophy

Francis Bacon, the State, and the Reform of Natural Philosophy

JULIAN MARTIN

UNIVERSITY OF ALBERTA

CAMBRIDGE UNIVERSITY PRESS

CAMBRIDGE

NEW YORK PORT CHESTER

MELBOURNE SYDNEY

CAMBRIDGE UNIVERSITY PRESS
Cambridge, New York, Melbourne, Madrid, Cape Town, Singapore, São Paulo

Cambridge University Press
The Edinburgh Building, Cambridge CB2 2RU, UK

Published in the United States of America by Cambridge University Press, New York

www.cambridge.org
Information on this title: www.cambridge.org/9780521382496

First published 1992
This digitally printed first paperback version 2007

A catalogue record for this publication is available from the British Library

Library of Congress Cataloguing in Publication data

Martin, Julian
Francis Bacon, the state, and the reform of natural philosophy / Julian Martin.
p. cm.
Includes bibliographical references and index.
ISBN 0-521-38249-1 (hard)
1. Bacon, Francis, 1561–1626. 2. Science – Philosophy – History.
3. Political science – Philosophy – History. I. Title. B1198.M36 1991
192 – dc20 90-24242 CIP

ISBN-13 978-0-521-38249-6 hardback
ISBN-10 0-521-38249-1 hardback

ISBN-13 978-0-521-03566-8 paperback
ISBN-10 0-521-03566-X paperback

For Geneva

Contents

Acknowledgements

This book is the product of my interests in the natural philosophy and the politics of early modern Europe and, more specifically, of my desire to integrate the history of natural philosophy with political and legal history in the hope of enriching our understanding of the natural philosophy of this era. My concerns, therefore, may be seen as 'interdisciplinary' ones, and they reflect the wide-ranging character of my own training.

I had not intended to pursue an 'interdisciplinary approach' when I became a student of history; it has been the consequence of the privilege of acquaintance with many learned scholars with very different interests. My father, and then Robert Schuler, Janos Bak, Steven Straker, the late Gerhardt Benecke, Brian Levack, Myron Gutmann, Guy Fitch Lytle, Sabine McCormach, Robert Palter and Adrian Wilson have all taught me more, and better, than they can have known then from appearances. To Andrew Cunningham, my Cambridge supervisor, go my very warmest thanks: without his encouragement and attention, it is unlikely the doctoral thesis which underlies this study would have been attempted, let alone completed.

I benefited immeasurably from the scholarly companionship offered me at Cambridge by Simon Deakin, Richard Grove, Ian Harris, Robert Iliffe, Adrian Johns, Robert Kilpatrick, James Moore, Iwan Morus, Simon Schaffer, Andrew Warwick, Alison Winter and Brian Wormald – the patience of all of whom has been often taxed by talk of Francis Bacon. Laurence Brockliss, Andrew Cunningham, Sir Geoffrey Elton, Lisa Jardine, Nicholas Jardine, Brian Levack, Graham Rees and Steven Shapin have all done their level best in suggesting ways to improve the quality and precision of this book. Having had the example and advice of such talented scholars and friends as these to draw upon, any failings in this study are due to an irremedial dullness of wit. I thank also the Wellcome Trustees for their generosity in granting me a Ph.D. Training Scholarship, the President and Fellows of Clare Hall, Cambridge, for the honour of a Research Fellowship and their conversation, and my equally affable colleagues in the Department of History at the University of Alberta. To my parents I owe great debts, yet I owe even greater ones to Geneva Moore, my wife, to whom this book is dedicated.

A note on the text

Nearly all references here to the writings of Francis Bacon are to the great Victorian editions of James Spedding, Robert Leslie Ellis and Douglas Denon Heath, *The Works of Francis Bacon*, 7 vols. (London, 1857–61) or to James Spedding, *The Letters and the Life of Francis Bacon*, 7 vols. (London, 1861–74). These titles are abbreviated as *Works* and *Letters*.

Because the *Works* incorporate many translations of Bacon's Latin treatises, notably in Volumes IV and V, references to these are included in parentheses; for example: 'Praefatio', *Instauratio magna* (1620), *Works*, I, p. 132 (IV, p. 21). The translations in the *Works* are by many hands (see J. Spedding, 'History and Plan of this Edition', *Works*, I), and I have amended several of these. Despite its textual inadequacies and oddity of arrangement, this edition remains the standard – at least until the completion of a critical edition of Bacon's works being prepared under the general editorship of Lisa Jardine and Graham Rees.

Many commentators conflate the *Works* and *Letters*, offering citations to *The Complete Works* or, say, to *Works*, Vol. XII, and the like: a small enough distraction, were it not for the fact that in America the seven-volume edition of *The Works* edited by Spedding, Ellis and Heath was issued in *fifteen* volumes. Considerable difficulties in trans-Atlantic scholarly communication have been the result. I append a table of comparisons.

Introduction

This study attempts to provide a historical explanation of the origin and nature of the natural philosophy of Francis Bacon. It is not a critical assessment of the worth of his proposals, in the manner characteristic of philosophers of science, nor is it concerned to situate Bacon with respect to one or more traditions of philosophy in the Renaissance – a venerable strategy in the historiography of ideas, which implies location itself to be much the same as explanation. It is, instead, rigorously local in its attentions. Its premise is that the origin and nature of Bacon's natural philosophy can be explained satisfactorily only with careful attention to *his* context. *In jure non remota causa, sed proxima spectatur*, Bacon remarked of the law, and the maxim is useful in historical practice, too. This study looks to the concrete and the proximate, and necessarily begins with a strong biographical element.

'He was a great reader of books', Bacon's chaplain recalled, 'yet he had not his knowledge from books alone, but from some grounds and notions within himself.'[1] A truly satisfying historical explanation of Francis Bacon's natural philosophy must include an answer to why and how it was that Bacon became involved in writing natural philosophy. Only by seeking out these 'grounds and notions' can we fully appreciate why he wrote it in the particular (and peculiar) manner in which he did. This study of Bacon, I believe, is entirely novel – both in its guiding assumptions and in the interpretation of his natural philosophy that rests upon them. These are bold claims, especially since a vast amount of commentary has been devoted to Bacon since his death in 1626, but I trust the reader will be persuaded by them.

Modern scholarly interest in Francis Bacon typically has been divided between three topics: his significance for the history of modern science; his role in Jacobean politics; and his prose style itself. The first has been the concern of philosophers, historians of the sciences and historians of ideas.[2] The second has been the concern of political and constitutional historians.[3] The third has been the preserve of scholars of English literature and rhetorics.[4] Such wide attention reflects the fact that Bacon not only wrote celebrated works of natural philosophy, but that he was a principal figure in the higher reaches of Jacobean government, and the author of a prose so striking that he is regarded often as the major prose stylist of the very decades when English prose was becoming a literary medium in its own right.

Yet the most arresting and continuing feature of Baconian scholarship is that it has produced several Francis Bacons, none of whom significantly overlap. Given the highly specialised character of modern academic training and organisation, this state of affairs is hardly surprising. To my knowledge, there has been almost no interest in reintegrating these partial images.[5] Historians and philosophers of the sciences, for example, have usually considered Bacon's career as a lawyer and statesman to provide little of value for their explanations of his natural philosophy. Indeed, his engagement in public life has often been seen as an embarrassment, and something which repeatedly distracted him from the pursuit of philosophy.[6] Political historians have taken the opposite view, regarding his concerns for philosophy as providing little of value for their explanations of his political career, beyond repeatedly distracting him from the pursuit of the law (in which his abilities are commonly acknowledged) and plausibly contributing to contemporary suspicions of his political acumen and to his inability to amass lasting political capital.[7] By contrast, this study insists that Bacon's legal and political career was crucial in the creation of his natural philosophy and that his natural philosophy cannot be separated from his political ambitions. It is not merely a descriptive statement but an explanatory one to say that his was, in fact, the natural philosophy of a late-Elizabethan statesman.

This is not to suggest that other English statesmen of Bacon's day saw much that was familiar in his natural philosophy; this was clearly not the case, and Bacon knew it. Yet he always regarded himself as an English statesman, rather than as a philosopher *per se*, and he always believed his philosophy was a contribution to the advancement of the English state. Bacon was dedicated to enhancing the powers of the monarchical state, and his proposals for the reform of the law and the apparatus of governance on the one hand and the ones for the reform of natural philosophy on the other were devoted to this single end. Furthermore, these apparently different schemes shared the same structures, the same techniques and the same terminology. They were, after all, the products of the same mind.

The arrangement of this book flows from these considerations. Chapters 1 and 2 address the question of why and how it was that Bacon came to believe problems of knowledge were a proper part of a statesman's concerns. This involves first describing the specific political tradition in which young Francis Bacon was reared and to which he fully committed himself in his late twenties, and then showing how the circumstances of the later 1580s turned his attentions to problems of knowledge. Many Tudor historians will be familiar with the major contours of the political narrative shaping these chapters, yet its provision is a signal departure from previous studies of Bacon's natural philosophy, which have taken his interest in natural philosophy for granted. For the reasons mentioned above, such a starting place is a necessity: why an ambitious scion of a family of elite Crown servants, trained and expecting to follow their careers, should have pondered deeply upon philosophical matters is, to say the least, an issue worthy of close examination. Chapter 3 addresses the question of why and how it was that Bacon should have become concerned with natural philosophy in particular. His earliest writings about natural philosophy manifest his reactions to English developments he regarded as politically dangerous and a serious challenge to the stability of the state. From the early 1590s, it was his opinion that natural philosophy could be refashioned into a splendid support for the Crown. When compared to the tantalising complexity of the writings of his later years, Bacon's tracts and letters of the 1590s are usually passed over as if they were mere *juvenalia* – an odd thought, since he was forty-two when Queen Elizabeth died in 1603. Simply because it was during King James' reign that Bacon issued books and letters in some of which we find his most extended discussions about a reformed law, a reformed state and a reformed natural philosophy, it does not mean we should look to the years in which his most famous pieces were written for the inspiration and perspectives that underpinned them. In Chapter 4, I discuss the purposes, structures, and some of the procedures of Elizabethan common law and the common lawyers, and then describe Bacon's state security employments in the 1590s to illustrate both his familiarity with the law and his characteristic point of view about service to the state. Chapter 5 explores Bacon's central ambition: to reform the apparatus of governance so as to create what he believed were the conditions for an imperial monarchy. Sweeping legal reforms were a major part of this plan. The polity of *New Atlantis*, and that of 'Solomon's House' within it, are examined here as well, for here, too, is illustrated Bacon's vision of an imperial state. Equipped now with appropriate explanatory tools to do so, in Chapter 6 Bacon's scheme for a reformed natural philosophy is examined as a whole: its aims, its bureaucratic methods and its procedures for 'discovery'. Bacon's prescriptions for the reform of particular sciences, the relation of his natural histories to his civil histories, his views on revealed knowledge, on the working of the human mind, and on particulate motion – to mention but a few important topics – I propose to discuss in detail elsewhere. I believe that the main points of such discussions are derivable from the argument found here, and that a general historical framework for explaining Bacon's philosophy is deserving of separate and prior treatment. What this study describes is the creation and the major components of an audacious programme for the reform of the state. Bacon's natural philosophy was a subordinate part of this programme and cannot be understood adequately in isolation from it. His was a natural philosophy made appropriate to a centralising monarchy.

In recent decades, Bacon's significance for the subsequent history of the sciences has been increasingly played down as we rid our historiography of various nineteenth-century assumptions. It is no longer thought, for example, that he was the prophet of 'modern scientific method', or that the Royal Society of London in the seventeenth century can be characterised as 'Baconian'.[8] True, it is readily acknowledged that Bacon was a figure of great importance in the eyes of seventeenth- and eighteenth- century natural philosophers, but the usual conclusion derived from this fact – that Bacon was merely an inspirational figure – works to reinforce a denial of a substantial place for him any longer in the history of the sciences. Yet Bacon need not be rendered marginal to our histories; looked at with hindsight, we can see that he indeed had an impact upon the practice and substance of the sciences, namely, by providing a persuasive explanation of the civil purposes of natural philosophy, and a persuasive model of how research into the natural world could be conducted and how the investigators could be organised. We may not think highly of Bacon's prescriptions for, say, the conduct of experiments and for the discovery of the principles of nature, nor admire the ends to which he dedicated his labours, but the institutionalisation of the sciences and their practitioners in the seventeenth and eighteenth centuries owes much to him. Moreover, Bacon's insistence that 'knowledge is power' is, I believe, best understood as meaning that knowledge should be harnessed so as to augment the powers of the state. Perhaps more than anything else, this arresting thought is Bacon's most enduring legacy. In the seventeenth century, this Baconian idea was a crucial issue, one which was most explicit among natural philosophers and their patrons, and one which was overtly political. What was the civil role of natural philosophy and of natural philosophers?[9] Just as we can see that a central theme in early modern European history was the struggle for civil power between central governors, local lords and private gentlemen, so, too, there was a struggle about the ownership, organisation, generation and applications of knowledge about the natural world. Was it, for instance, to be conducted by (and for) the state, and within official academies, or was it to be conducted by (and for) private gentlemen and scholars, and by means of individual inquiry? Bacon's aphorism encapsulated his unequivocal answer: knowledge should be a department of the state. In this study, I shall show why and how he came to this belief, and how his natural philosophy sprang from it.

1

A statesman's responsibility

> To crown all, as she was most fortunate in all that belonged to herself, so was she in the virtue of her ministers. For she had such men about her as perhaps this island did not produce before. Yet God, when favouring kings, also arouses and enhances the spirits of their ministers.
>
> *In felicem memoriam Elizabethae Angliae Reginae* (1608), *Works*, VI, p. 296

In 1590, Francis Bacon was thirty years old, and deeply engaged in the vocation which he always had desired for himself: *statesmanship*. Explaining what he understood by statesmanship is the aim of this preliminary chapter, which concerns the establishment in the highest reaches of mid-Tudor government of men with a particular conception of the role and purpose of a royal councillor: namely, the notion of a 'commonwealth' statesman. Just why we should interest ourselves in such a theme is not immediately obvious: certainly, no previous study of Bacon's philosophy has done so. Yet these men included members of Bacon's family and kin, devoted to serving the Tudor dynasty and fully accepting a 'commonwealth' notion of themselves and their work. It was, after all, in their company and amid their values that Francis Bacon grew up.

A programme for statecraft, which incorporated a variety of reforming proposals into long-term state planning (as we might call it) had been learned by Bacon's father, his uncles and their colleagues when junior officials of the Crown under Thomas Cromwell and his successors. The commoners who became principal councillors and officials of the young Queen Elizabeth had received their introduction to royal office during the late 1530s and the 1540s. They were employed in the restructured (or newly created) courts and offices of the central government and they became deeply involved in the detailed business of administering the host of new laws passed by the Reformation parliament and those which followed it.

The sweeping changes in England during the 1530s were due principally to the efforts of Thomas Cromwell, the king's great minister, whose vision of a reformed and self-reliant English commonwealth shaped the legislation he prepared and steered through the parliaments of the 1530s. Cromwell was guided by deep convictions about what the realm of England should be, and it is accurate therefore to speak of Cromwell's reform 'programme', rather than regard his achievements in the governance of the state as resulting from a series of piece-meal responses to his current circumstances. Because he dominated the royal administration during the 1530s – accumulating great patronage power, and using it to ensure his confidants and lieutenants controlled various parts of the machinery of governance – the junior men then offered Crown service were those considered 'sound' and of like minds by their Cromwellian superiors.

These younger men survived Cromwell's fall in 1540, and they slowly and warily advanced themselves in office during the conservative retrenchment of Henry VIII's declining years, the chaotic politicking of Edward VI's short reign and the religious reaction under Queen Mary. During Cromwell's administration, and following his example, they had learned to be careful reformers: just what sort of particular policies for the enhancement of the 'commonwealth' were capable of becoming successful legislation? These lessons in statecraft were reinforced while serving the Crown during the uncertainties of the regimes which followed after 1540.

When Elizabeth ascended her father's throne in 1558, many of her close advisors and councillors were chosen from this group of (now middle-aged) professional Crown servants, who had retained their youthful convictions about the reforming purpose of their work in central government. Among these men were Francis Bacon's father, Sir Nicholas Bacon, and two of his uncles, Sir Thomas Gresham and Sir William Cecil. During the first decades of Queen Elizabeth's reign, many of the domestic policies of her government were formed and executed by these men, and in accordance with the vision of a reformed commonwealth – and the means of achieving it – which Cromwell had shared with his confidants and their young officials.

Commentators often suggest, casually, that the early sixteenth century in England was a 'time of reform' – that 'reform' was 'in the air' – and they direct our attention forthwith to the publications of 'humanists', men such as Erasmus, Thomas More and Thomas Elyot. This begs a series of questions for the historian, the principal one being the pressing matter of displaying concrete causal relationships between a 'climate of ideas' and known practical action. A related issue centres upon the meaning(s) of the word 'reform'. The noun 'reform', the *Oxford English Dictionary* tells us, first occurred in the eighteenth century, and the senses of the verb 'to reform' which were most common among the early Tudors were 'to renew', 'to form again' and 'to restore'. None of these carry the idea of sweeping changes for the better which we now automatically associate with the verb 'reform'. It is unlikely, for instance, that Erasmus, More or Elyot spoke of themselves as 'reformers'; the first published instance of the word occurred in 1548 – when all three men were dead. It is more appropriate to say of such men that they were scholarly men who wanted 'to restore' rather than 'to reform'.

In general, English humanists were moralists: pious, intellectual and very conservative. Educated in Italy in the newly discovered 'pure' languages of classical Rome and Athens and in the techniques of rigorous textual criticism, they returned to England burning with desire both to teach the 'new learning' in the schools and universities and to toss out the 'old learning' – namely, the scholastic disputation and commentary upon time-honoured, but (in their eyes) textually deficient works of philosophy, theology and Scripture itself. They had profoundly devout aims and they shared a belief that the new scholarship would improve scholars' understanding of Scripture (and thus theology) and would help also to increase the religious fervour of the young clerics whom they taught and who were to minister to the people.[1]

By 1510, humanist scholars had extended their operations from the universities to London and the royal court, where they received encouragement and patronage from Lady Margaret Beaufort, the grandmother of King Henry VIII, from Henry's queen, Catherine of Aragon, and from Lord Mountjoy, to name but the most prominent friends of the 'new learning'. At Henry's court, the humanists were particularly involved with the 'proper' (i.e., humanist) education of noble children, for they argued, ingeniously, that not just scholars would benefit from their new programme of classical learning but the future ruling classes as well. Rigorous training in classical Latin and Greek (and often in Hebrew), studying and translating the Scriptures in their ancient form, and reading classical secular authors (especially the historians and the moralists) were championed as the means to enhance the moral behaviour and intellectual profundity of those who were the 'natural' leaders of English society and its government.[2]

Some scholarly training of noble children in classical languages and literature, educational reforms for the clergy, and attempts to revive a spirituality dissipated amid penances, indulgences and paid-up masses that rendered Christianity more legalistic than heart-felt: English humanists did these worthy things, but it would be a mistake to regard them as 'reformers' on many fronts. The satirical, even bitter, attacks by humanists on the state of the Church or secular society which come quickly to mind – publications such as Erasmus' *Praise of Folly* or More's *Utopia* – were highly conservative in spirit and their authors more concerned with the proper shepherding of the Christian flock, the salvation of souls and the conservation of the God-given order in the world than ever they were with the alteration of worldly institutions or the adoption of new ones.[3]

This is not to suggest that educated men did not criticise the state of affairs in the realm and desire alterations in whatever it was that rankled them; they did, but we should be wary of regarding their protestations as exemplifying a general mood for 'reform' as we understand it. Even if the humanists' books and sermons further sensitised a gentleman to the worldly ills and evils that surrounded him, neither books nor sermons were the instruments by which he might achieve the restoration of an earthly goodness or even the elimination of a specific ill: in Tudor England the only such instrument was legislation.

Legislation in England meant either the use of royal proclamations or, more commonly, the passage of statutes by a parliament.[4] Because of just who it was that composed the membership of parliaments, it was specific and concrete grievances – usually those affecting a locality or a particular group – which stood the greatest chance of being translated into law. Among the sorts of grievances that did receive considerable attention in early Tudor parliaments were complaints about particular conditions in (what we would call) agriculture and the economy. Debates about depopulation, enclosures, vagrancy, corn prices, weights and measures, wages, guilds and craft practices, the textile trade and the like were common fare in parliament for the simple reason that parliaments were composed largely of the great landowners whose concerns these naturally were.

It is difficult to conceive of the Commons and Lords promoting grand changes in the structure or habits of English farming and trade, let alone alterations in major institutions; not only would they be unlikely to endanger their present advantages but they did not regard such things as their duty. Proposals for major changes (i.e., sweeping, general legislation) were the prerogative of the Crown itself, and during the first thirty years of the century there is little to indicate that the king and his councillors were interested in this sort of legislation – despite the humanists at the court and their moralising. This situation changed suddenly with the extraordinary crisis prompted by King Henry VIII's demand for legitimate divorce from his queen and with his decision to accept the counsels of Thomas Cromwell.

Only with Thomas Cromwell's ascendancy can we begin to speak about 'reform' in the sense in which we use the word. The course of Cromwell's rise to great power and the changes which he instituted in the governance of Church and state has been recovered in detail by Professor Sir Geoffrey Elton. What is (for our purposes) one of Professor Elton's most important arguments about Thomas Cromwell is sometimes overlooked: namely, that along with other and more pressing campaigns, Cromwell had a programme by which he hoped to improve the 'commonweal' (i.e., the common 'good', 'benefit' or 'welfare' of the people). In many themes and specific targets his programme seemed to echo (not surprisingly) the grievances voiced during previous decades by a wide variety of men but it was *Cromwell's* plan which was instituted in the 1530s, and not that of anyone else.[5] This may seem an obvious point, but the considerable modern interest in the Tudor discussions of the 'commonwealth' often has been combined with a simple equation between the humanists, their writings and the direction the legislation of the 1530s actually followed. Some scholars have tried to explain the legislation by looking for 'influences' upon its principal agents amid the humanist tracts,[6] others by arguing that such tracts are evidence of a 'commonwealth party' in government during Henry's reign and after,[7] and yet others by claiming that Cromwell and his friends were fundamentally and essentially themselves humanists.[8] These arguments have not withstood close scrutiny.[9] Certainly, Cromwell was well acquainted with the humanists Thomas More and John Rastell from the early 1520s, and his important place in Cardinal Wolsey's household brought contact with many other scholarly men both English and foreign, but Cromwell and his programme of 'commonweal' reforms gained inspiration from sources other than humanism.

Cromwell had travelled on the Continent between 1500 and 1512, working in Italy and the Netherlands. Between *c.* 1503 and *c.* 1508, he was employed first by the Frescobaldi family of Florentine bankers and then by a Venetian merchant. He later left Italy for the Netherlands and Antwerp (the principal trading 'vent' for English traders) where for four or five years he was an advisor to various English merchants trading between London and the Continent. By 1512 he had returned to London. Cromwell probably gained his life-long fascination with the problems of the best way to order a state when in Italy, where mounting threats to the independence of the republics of Florence and Venice produced much argument there about the nature and uses of political sovereignty in a state.[10]

Alongside his fascination with Italian political arguments, we must count Cromwell's religious convictions as a source for inspiration. His own doctrinal position is difficult to uncover, yet – during the 1530s – Cromwell certainly became more Protestant than his king. Cromwell was genuinely passionate about the Bible. While on a trip to Rome in 1517–18, he is thought to have learned by heart Erasmus' translation of the New Testament, and during table talk he was able to quote freely from throughout the Bible.

> In Cromwell's case, [Bible-worship] supplied one of the driving forces to an essentially political temperament, the principled undertone and transcendental justification of labours that concentrated upon reforming the earthly existence of men by reconstructing the state.[11]

But why should Cromwell have decided to reform 'the earthly existence of men' and why should he have decided to do so by 'reconstructing the state'? Yet the extent of his achievement, and the ongoing controversy over it, obscure the truth that it was the means to an end, and that the *purposes* for Cromwell's labours were more radical than what he actually brought about. 'Reforming the earthly existence of men' was anything but an orthodox aim for early sixteenth-century politicians – the very idea was extraordinary. Humanism cannot provide a sufficient explanation for Cromwell's purpose and the same can be said of his Bible-piety. A completely satisfying answer about the formation of Cromwell's motives may remain beyond our resources but it is worth looking into Cromwell's own past and particularly into his years in northern Italy, as a strong claim can be made that he became enamoured with the state envisioned by Marsiglio (Marsilius) of Padua (1275–1342).[12] Marsiglio's manuscript *Defensor pacis* often had been recopied and it was widely known in northern Italy by 1500.[13] We know that Cromwell in the 1530s displayed great admiration for Marsiglio and even ordered the translation of *Defensor pacis* into English in 1535.[14] What can we divine from this?

When Cromwell was living in Florence and then Venice, northern Italy had been a battleground for more than a decade. The successful invasion by the French king in 1494 had undermined the precarious balance of the Italian states and had precipitated a constant grind of shifting alliances, wars, revolutions and counter-revolutions. One upshot of the political and economic turmoil was a frenzy of concern about how best to order and preserve a state. Florence, for example, had oscillated between popular democracy and tyranny after 1494 and, at least from 1502 until 1506, a circle of Florentine aristocrats and their scholarly friends met regularly and openly in the Rucellai Gardens to discuss how they should organise the state if and when they could gain power.[15] These men wanted an aristocratic regime with a monarchical head to rule Florence, and they chose this arrangement not only because it benefited them but because they thought it was how Venice was organised, and the stability and prosperity of The Most Serene Republic were the envy of war-torn Italy.[16] At just this time, Cromwell was employed in Florence by the aristocratic Frescobaldi. Although there is no shred of evidence linking him with the Rucellai Garden circles, he could hardly have been unaware of them and the topics they debated there. It once was believed simply that Cromwell was inspired by Machiavelli, who was writing throughout the period Cromwell was in Italy,[17] but both in Florence and in Venice, political argument and theorising repeatedly recalled the work of Marsiglio, and it now seems certain that Cromwell met men who introduced him to the works of Marsiglio and those who had followed his lead in arguments about the nature of the state and how to preserve it in prosperity and peace.

A state, Marsiglio had argued (echoing Aristotle), was a construction of human reason with a finite purpose which was wholly secular and earthly: the maximum 'civil happiness' of the citizenry. Just like a human body, the state was composed of a variety of differing 'parts', each with complementary functions, and 'tranquillity' resulted from the 'best disposition' of these parts.[18] 'Tranquillity' was for Marsiglio not simply the eventual result of 'right reason' or 'virtue' in the citizens, but it was the product of safety, lawfulness and economic well-being. The virtuousness of the citizens is indeed the final end of the state but Marsiglio concentrated his attention upon its essential preconditions, which are material and mundane and are accomplished through rational calculation and legislative action. The ruler of this very unmedieval state is the Defender of the 'peace', and his responsibilities are nothing more than achieving and preserving this prosperous state of earthly affairs.[19] On the very first page of *Defensor pacis*, he announced the proposition which may well have inspired Thomas Cromwell's aim of reforming the earthly existence of men:

> Tranquillity, *wherein people prosper and the welfare of nations is preserved*, must certainly be desirable to every state. For *it is the noble mother of the good arts*. Permitting the steady increase of the race of mortals, *it extends their powers and enhances their customs*.[20]

Cromwell's exposure to the precariousness of the Italian states, to political activists in Florence and Venice, and to Marsiglian arguments for promoting the economic well-being of the people was followed by several years as an advisor to English merchants in the great entrepot of Antwerp, England's sole significant trading link with the Continent until the 1550s. In the 1530s, Cromwell resisted several parliamentary attempts to restrict the trade there, and we may surmise that it was his work in Antwerp that consolidated a conviction of the critical role played by this market in sustaining England's textile-based economy.[21]

Marsiglio's doctrines about the sovereignty of the secular state and the illegitimacy of the temporal pretensions of the papacy were attractive cannon-fodder for Cromwell in the propaganda war fought over his king's divorce, but Marsiglio was of much greater importance to Cromwell than mere propaganda; Cromwell's radical vision of the reformed realm of England was inspired by Marsiglio. The preambles to Cromwell's great statutes in the 1530s support this conclusion, and none more strongly than the language of the 1533 Act of Appeals, which encapsulated the thrust of Marsiglio's arguments:

> Where by divers sundry old authentic histories and chronicles it is manifestly declared and expressed that this realm of England is an empire, and so hath been accepted in the world, governed by one supreme head and king having the dignity and royal estate of the imperial crown of the same, unto whom a body politic, compact of all sorts and degrees of people divided in terms and by names of spirituality and temporality, be bounden and owe to bear next to God a natural and humble obedience ...[22]

Although they all are manifestations of his single vision and grand purpose, here we especially are concerned with Cromwell's plans for the English 'commonwealth', not with his convictions about the constitutional character of the realm nor with the entirety of his reforms in governance. Any study of Cromwell's achievement, whether concerned with the subjugation of the Church, the destruction of independent jurisdictions in England, the restructuring of the central administration or with the 'commonweal' reforms, must be 'a study of parliamentary legislation – planned, attempted, achieved or lost'.[23] Turning his beliefs into a programme of action and translating this into successful legislation required Cromwell to encourage and to organise trustworthy advisors and publicists,[24] to mobilise the government's supporters and officials,[25] and then to recruit for various administrative courts the minor officials who would monitor compliance with the new legislation and adjudicate firmly in the squabbles it would provoke.

Cromwell's 'commonweal' reforms were of two broad sorts: those concerned with agricultural and economic issues, and those dealing with the failings of the law.[26] From 1532, when Cromwell achieved the dominant position in the king's council, great numbers of 'commonweal' bills entered parliament: most were approved as new statutes either directly or upon reintroduction in subsequent sessions.[27] The agricultural and economic bills tried to stop the rash of enclosures and the reduction of tillage, to solve 'vagrancy', to encourage urban rebuilding, to find ways for looking after the poor and sick, to strengthen the weak domestic textile industry and to promote a healthy export trade for manufactured goods. The law, its administration, the confusing number of competing jurisdictions, the expensive delays and the wide-spread suspicion that lawyers were greedy had long been topics for bitter complaining in Tudor parliaments, yet Cromwell's achievements here were substantial.[28] He was able to reduce the privileges of benefit of clergy and of the right of sanctuary, both of which often frustrated the course of justice and the pursuit of criminals,[29] but much more was done in matters of the secular law. A series of minor acts were passed against particular abuses in legal procedures and for the strengthening of the administration of the laws;[30] the Statute of Uses (1536) and the Statute of Wills (1540) 'transformed the principles of landownership';[31] procedures in the Courts of Star Chamber, Requests and Chancery were stabilised; and the new Courts of Augmentations and Wards were created.[32] One of the immediate boons of Cromwell's attention to the great courts (and his creation of new ones) was speedier and more efficient justice. It also meant central governance needed new officials to staff the new and improved courts.

The expansion of the numbers involved in central administration occurred principally on behalf of the new revenue courts: the Court of General Surveyors, of Wards, of First Fruits and Tenths, and the great Court of Augmentations. These creations were designed to administer and audit the vast lands, buildings, fees and incomes which recently had become the property of the Crown.[33] Despite the enormous increase in the Crown's revenues and the administrative duties which went with it, the increase in the numbers of central government officials was measurable in dozens of men. The greatest number of posts available under the new courts were local ones – looking after the affairs of a specific estate, park or wood, or supervising a group of properties in a particular district – and commonly the men who did these jobs were the ones who had done them for the previous owners. The regional and central superstructures were indeed new, and the men who won the demanding posts of regional overseers, auditors, general receivers and legal councillors overwhelmingly were university-educated commoners, most of whom were young and many of whom were lawyers, nibbling ambitiously at the fringes of the royal court and the royal administration. Most of these young beneficiaries of Cromwell's administrative reforms, being political ciphers, remained employed in the central revenue courts after his fall. Many served the Crown with increasing competence during the troubled last years of Henry VIII and the regimes of Somerset and Northumberland. A handful, who had achieved prominence in office during the reign of Edward VI and then discreetly curtailed their ambitions (or simply retired) under the Catholic Queen Mary, were rewarded upon the accession of Queen Elizabeth with positions of great responsibility in the realm.

Typically, the commoners who entered Elizabeth's inner councils in 1558 were experts in the detailed operation of the central machinery of governance. Their known administrative competence and energy, their political and religious discretions and their loyalty to the Tudor dynasty had won them the confidence of the new queen.[34] They also had shown their continuing commitment to the reforms begun in the 1530s, to the established Church and to the 'commonweal' programme. Richard Sackville, Thomas Smith, Thomas Gresham, Walter Mildmay, William Cecil and Nicholas Bacon shared much with their new queen. Taken as a group, these men can be described as canny 'survivors' – which is not to criticise them but only to highlight the sensitivity of their political antennae, and their attitude to the Crown and to their own roles as administrators. In matters of religion, they were conservative, 'orthodox' Protestants rather than passionately evangelical, and willing to accept the dictates of the Supreme Governor of the Church rather than follow their own consciences. In other matters of state, they were determined to give sincere and well-informed counsel to their queen (even when it was undesired) and then to execute her commands as speedily as possible. Elizabeth's famous remark to

Cecil upon appointing him Secretary of State might be applied equally to those other commoners she appointed in 1558:

> This judgement I have of you, that you will not be corrupted by any manner of gift and that you will be faithful to the state, and that without respect of my private will you will give me the counsel which you think best.[35]

One element in being 'faithful to the state', or so these newly appointed statesmen believed, was the duty to reform the English 'commonweal', a belief and a programme which they had imbibed under Cromwell and his successors. This commitment is easily recognised in Thomas Smith, who in 1549, while Secretary of State, had written a manuscript entitled 'A Discourse of this Commonweal of This Realm of England', in which he persuasively analysed the economic ills of the nation and proposed several legislative cures for them.[36] Smith's manuscript was printed in 1565, nearly twenty years after its composition, under a regime completely sympathetic to his arguments.

Smith believed the present dearth, the 'decay of Trade' and the hunger of the poor were avoidable, and he condemned as central causes of the economic crisis the recent debasements of the currency and the general – and increasing – reliance upon foreign manufactured goods.[37] The reliance upon foreign commodities and the circulation of bad coin produced a nationwide poverty and led to price rises and to the decay of native trades and manufactures. Smith advocated a return to good coinage, and the stimulation of manufactures by taxing imports and by bringing in skilled foreign craftsmen to establish industries which England lacked.[38] By these governmental measures, he thought not only to increase England's self-reliance in commodities but to export goods as well:

> We should not only have enough of such things [manufactured goods] to serve our realm and save an infinite treasure that goes now over for many of the same, but also might spare of such things ready wrought to be sold over, whereby we should fetch again other necessary commodities or treasures.[39]

Smith hoped his 'Discourse of the Commonweal' would persuade the Protector Somerset and the Privy Councillors (alarmed by agrarian uprisings and the disorders of the economy) to pass the reforms he advocated, but it failed to do so. In fact, Smith fell into disfavour shortly after writing his tract and then slid into political eclipse.[40] Nonetheless, his opinions and reform plans (and probably his manuscript as well) became well known among his friends,[41] and the 'commonwealth' reforms of Queen Elizabeth's early years were strikingly similar to those which Smith unsuccessfully voiced in 1549. Two principal reasons why this was so were his two close friends, Sir Nicholas Bacon, now the Lord Keeper of the Great Seal, and Sir William Cecil, the Principal Secretary of State.[42] Bacon and Cecil were largely responsible for preparing the new government's legislative programme, and during the very difficult and dangerous early years – the 'testing time' of the new regime[43] – while the Privy Council faced the immediate problems of foreign war, of rival claimants to the throne, the succession problem, and the clash of conservatives and radicals over the queen's religious settlement, Bacon and Cecil believed it important to proceed with economic and legal reforms.[44] Both sorts of reforms and the purposes for them clearly bore the stamp of Thomas Cromwell.

Cecil was especially impressed with the need to decrease reliance on foreign goods and to encourage native industries. Throughout much of his long career as Elizabeth's principal councillor, he pursued a programme designed to increase English wealth, self-sufficiency and employment from manufacturing. This programme had several tactics. One was to put restrictions upon the sale of foreign-made goods, especially clothing (which hurt the principal native industry).[45] Another tactic was to restrict severely the export of raw materials, which might return as finished goods for sale and put English coin into the pockets of foreigners.[46] Cecil also strove to promote the English shipping industry and the merchant navy; one of his methods was to declare Wednesdays, as well as the usual Fridays, as 'fish days', and thereby sustain a strong native fishing fleet.[47] In addition to these legislative tactics, Cecil also relied upon the issue of letters patent for special economic 'projects'. At the beginning of the reign, Cecil adopted a Continental practice and began to grant patents of monopoly (protective industrial privileges) for a term of years to inventors or skilled craftsmen who seemed able to establish a new industry in England through their exploitation of a process, technique or machine then unknown or unused in England. By issuing these protections, Cecil hoped to nurture new industries, to increase employment, to achieve the reforms for which Smith had argued in his 'Discourse of the Commonweal' and, in short, to make England a more prosperous and secure nation.[48]

Sir Nicholas Bacon as well as Cecil had a principal place in forming the government's domestic programme. In his speeches to the parliaments, in his work on the Crown's revenues and on the recoinage, and especially in his reforms of the law, Bacon continually strove for the improvement of the 'commonweal' of England.

In late 1558, the Privy Councillors named a committee with Bacon as its head to investigate the state of the Crown's revenues and expenditures. Bacon in 1559 recommended setting up two committees to make a thorough examination of the Crown's finances. These committees, each entrusted with different aspects of the Crown's affairs, continued to report during the next few years.[49] The Privy Councillors' increasing knowledge of the true extent of the queen's financial obligations and of the details of her government's ordinary expenses allowed them to attack and reduce the Crown's expenditures successfully. In his last years, Henry VIII had diluted the gold and silver content in English coins, and although coins of higher quality were minted by Northumberland's government in 1552 and by Queen Mary, Henrician coins remained in circulation and they continued to cause havoc with prices and trade. From early 1559 until late 1560, Bacon assisted Cecil in evaluating the reports from various experts about the many problems of the recoinage.[50]

In January of 1559, opening Queen Elizabeth's first parliament, Bacon delivered a speech outlining the government's concerns and its expectations of the assembled Lords and Commons. He asked them to consider the establishment of a religious settlement, the reduction of the nation to civil order and lawfulness and 'the state and condicion of the realme, and the losses and decayes that have of late happened to the imperiall crowne thereof, and thereuppon to devise the best remedyes to supplie and relieve the same'.[51] At the close of this parliament, as its members were preparing to return to their 'countries', Bacon charged them to be diligent magistrates, and he particularly condemned as 'the root and feede of all injustice' the abuse of legal process by Justices of the Peace:

> Surely, surelye, it is true that thease be they that be subverters and perverters of all lawes and orders, yea, that make dayly the lawes that of thier owne nature be good to become instrumentes of all iniurye and mischiefe; thease indeede be they of whom such examples would be made as of the founders and maynteyners of all enormities, and theas be those whom, if yee cannot reforme for their greatness, yee ought heere to complayne of for their evillness.[52]

Bacon made some reforms in the administration of the great central Court of Chancery, but he remained especially concerned about the quality of royal justice available in the countryside.[53] His speech to parliament in 1559 was followed a few weeks later by a similar exhortation to the assembled Justices of Assize, and he repeated its main themes again and again to parliaments and the judges.[54] The Justices of the Peace were unpaid, indifferently trained and nearly always the principal landowners of the county; however much Bacon and the Privy Council wished them to be obedient and efficient local agents of the Crown's wishes, the energy and impartiality with which they dispensed royal justice never could be taken for granted.[55] Bacon ordered in 1561 and in 1562 the distribution to the JPs of an 'Abbreviate' of the current statutes; at least no justice could plead ignorance of which laws he was expected to enforce.[56] The Lord Keeper successfully pushed for a large reduction in the number of JPs. In 1562 the commissions of the peace – *libri pacis* – were greatly trimmed down: a salutary warning that good conduct was expected.[57] In 1563, during Elizabeth's second parliament, six statutes were created condemning a variety of abuses of justice and closing legal loopholes; we may assume by their form and from their national, rather than local, character that they were bills promoted by Bacon and the Privy Councillors.[58]

In his opening speech to the 1563 parliament, Bacon threatened the members with a system of visitation upon the JPs by 'experte and proved persons' every second or third year, and he emphasised the parallel of the ecclesiastical visitations of the clergy.[59] He obviously believed the JPs' role in the local community was much more than an after-the-fact retributive one. As the surviving state papers illustrate, Privy Councillors directed a flood of correspondence to individual JPs and county benches of JPs urging *activity* upon them, not merely swift execution of their commands about specific problems. JPs were expected to promote in the people *compliance* with the behaviour mandated by the huge number of statutes, royal proclamations and specific orders from the Privy Councillors. An effective Justice of the Peace (in the councillors' opinion) would constantly have been busy announcing what the laws were, and thereby teaching good (i.e., 'lawful') behaviour. The JPs, then, as administrative agents of the government had not only a duty to punish transgressors but they had a public ministry of sorts as well. William Lambarde said as much in 1585: 'The laws themselves be the *outward guides* and masters of our lives and manners.'[60]

These government programmes of encouragement for foreign craftsmen and new industries, of financial stock-taking and of intensive scrutiny into the administration of justice are the signatures of the professional administrators within the Privy Council, and the thoroughness with which these reforms were pursued suggests the degree to which the young queen relied upon Bacon and Cecil and their reform-minded colleagues in the shaping of her policies. These men had become Queen Elizabeth's councillors after a long apprenticeship under her predecessors. Their training in the administration of government provided both the professional skills and the political sense needed when deciding what government policies would be practicable for the nation they governed in the queen's name. In important addition to their practical skills, they had entered the service of the Crown during the 'Cromwellian' era and they accepted reform of that sort as part of the natural work of a statesman. In particular, Nicholas Bacon and his brother-in-law William Cecil retained the 'commonweal' beliefs of their younger days and they assumed that their duties as royal councillors included planning reforms of the nation's economy and the law. They accepted as a fundamental duty the responsibility for enhancing the 'commonweal' of England through their administrative and legislative labours. Bacon's speech opening the 1571 parliament reflected this conviction and it accurately captures many of the themes we have pursued here:

And what is peace? Is it not the richest and most wished for ornamente that pertaines to any publique weale? Is not peace the marke and ende that all good governmentes directes their actions unto? Nay, is there any benefitte, be it never soe greate, that a man may take the full commodytie of without the benefitte of peace, or is there any soe litle commodytie but throughe peace a man may have the whoall fruition of it? By this we generally and ioyfully possesse all, and without this generally and ioyfully possesse nothinge.[61]

2

The young statesman

> Since my birth and education had trained me for civil affairs, and being in opinions (since but a youth) sometimes shaken, and considering I owed my country something special not equally belonging to every other party, and expecting, if I obtained some honourable place in the state, a greater assistance of talent and diligence to carry out what I intended, I commended myself, with due modesty and ingenuousness, to such friends as possessed some influence.
>
> *De interpretatione naturae prooemium* (*c.* 1603), *Works*, III, p. 519

Francis Bacon's youth was spent in preparations for a life of royal service. Sir Nicholas carefully guided his formal education toward this end, and Francis seems always to have believed that to serve as a councillor was his vocation. While a young man, Francis was confronted with several differing conceptions of just what the role and the purpose of a statesman should be, yet by the late 1580s he had made his own decision about the responsibilities of a good statesman. His youthful experiences and the nature of his employments after 1586 made Bacon utterly convinced that the nature and problems *of knowledge and learning* were properly the concern of the statesman he wished to become.

Francis Bacon's formal education began under the watchful eye of his parents but we know almost no details of the schooling he received in his father's house. In his lessons he was accompanied by his brother Anthony and by several wards, mostly kinsmen, whom the Lord Keeper had taken into his household. He learned Latin grammar, the Bible, the prayers and religious exercises of the Church, and other primary school accomplishments that great men of Elizabethan England expected of their sons. Who tutored him is uncertain: probably the successive chaplains in the Bacon household led the children through their elementary lessons. We are told that Francis at an early age was seen to be very bright; the queen apparently called him 'the young Lord Keeper': she was amused by his precocious *gravitas*.[1]

In the springtime of 1573, when Anthony was fifteen and Francis just twelve, Sir Nicholas sent them to Trinity College, Cambridge, putting them under the personal tutelage of the Master, Dr John Whitgift.[2] Whitgift by the 1570s was the university's vice-chancellor and a highly controversial figure. His administrative skills, his gathering of collegiate powers into the hands of the Masters alone and his successful battle to quell religious nonconformists in the university had brought him the favourable attentions of the queen, of Burghley (the chancellor of the university) and, we must suppose, of the Lord Keeper as well. By 1573, and for the same reasons that won him the queen's support, Whitgift had gained the undying enmity of puritan activists. Lady Anne Bacon must have been grievously unhappy when her husband entrusted her sons to the hammer of the 'Godly'.[3] The Lord Keeper did not share his wife's religious predilections: if he had done so, his placing of the boys under Whitgift's personal care would remain inexplicable.[4]

Whitgift knew what his patrons wanted: vigilant protection for the training ground of the established Church and personal attention to the schooling of the boys sent to him by Burghley and other great men in the realm. He was personally ambitious, politically astute, and deeply suspicious of the passionately evangelical members of the university. We should expect Anthony and Francis Bacon heard a very great deal of the sort of Scriptural gloss Whitgift provided in a sermon before the queen in 1574:

> *A wise man continueth in his wisdom constant as the sun, but a fool is altered and changed even as the moon.* By a wise man he meaneth such as are guided with reason and knowledge; by a fool he understandeth, not those that lack the use of reason, and be void of wit, but those that follow affectation rather than reason, and are carried away with opinion and fancy, not with sure proofs and certain knowledge.[5]

Whitgift believed authoritative governance was necessary for the university and the Church, and that puritanism (or 'faction' as he called it) was a form of spiritual rebellion, corrosive of individual morals and a solvent to the fragile unity of the Anglican community. In the same year, 1574, he wrote:

> I do charge all men before God and his angels, as they will answer at the day of judgement, that under the pretence of zeal they seek not the spoil of the church; under the colour of perfection they work not confusion; under the cloak of simplicity they cover not pride, ambition, vain-glory, arrogancy; under the outward shew of godliness they nourish not contempt of magistrates, popularity, anabaptistry, and sundry other pernicious and pestilent errors. The Lord make us thankful for his infinite mercies and singular goodness bestowed upon us, in thus long continuing his gospel, preserving our most gracious and loving queen, and overthrowing all the conspiracies and devises that the devil hath hitherto invented to molest this state and church.[6]

Whitgift's young charges studied the usual curriculum: they read the set texts of the undergraduate exposure to the *trivium* (i.e., grammar, rhetoric, dialectic), the *quadrivium* (geometry, arithmetic, astronomy, music) and ethics, natural philosophy and metaphysics.[7] After slightly less than two years in Cambridge (plague closed the university from August 1574 until March of 1575), Francis and his brother returned to London, but not to their father's residence, York House. Instead, they were admitted into Gray's Inn to begin study of the common law, and they took rooms there which the Lord Keeper, as the most eminent member of Gray's, had reserved to his use. Sir Nicholas' sons were accorded the status of 'ancients' and were released from some of the regulations for 'continuance' which other novices had to obey.[8] However, as with all the novices ('inner barristers'), they followed the intense cycle of law readings, 'bolts', 'moots' and other exercises by which the senior members of the Inn trained the younger men in the arts of legal argument, providing them with the knowledge needed to become professional common lawyers if and when they were deemed ready.[9]

The barristers and law students were anything but isolated at their Inns; they were at the centre of the nation's capital and were involved intimately in its legal and political business, gossip and intrigues. Moreover, the London households of the great nobles and the officers of the Crown were hives of business activity, and members of the Inns were well acquainted with them all. For example, Leicester House (near Gray's Inn) was the home of Robert Dudley, the Earl of Leicester, and of his nephew, Sir Philip

Sidney, who encouraged visits from the young men of the Inns. As the sons of a Privy Councillor and England's most powerful judge, Anthony and Francis were welcome visitors there.

The Earl of Leicester had risen to great power in the 1560s. Leicester became Queen Elizabeth's favourite, and then a Privy Councillor (and an aristocratic counter-balance in the council to the administrator-professionals). He and William Cecil remained careful rivals and in competition for the queen's ear.[10] Leicester soon began to use his position to advance the cause of further religious reform in England. By the mid–1570s, he was the recognised political spearhead of the English Puritan movement, using his favoured position on the Privy Council and the offices and honours showered upon him by the queen and well-wishers to build a web of clients throughout the country. 'Godly' gentlemen, clerics, scholars and adventurers staffed his household. With his powers of appointment, Leicester advanced and protected Puritan clergy and laymen alike.[11]

Prominent among the members of Leicester's entourage was Sir Philip Sidney, a striking young courtier, a poet, a patron and correspondent of scholars, and greatly favoured by the queen and by several foreign princes. Sidney had become a champion of the international Protestant cause during the 1570s. He was a firm advocate of a Protestant foreign policy and desired a holy war against Spain.[12] Many hopes were pinned upon him, including those of both Leicester and his brother, the Earl of Warwick, who were childless; it was widely expected that when they died they would leave a very great deal to Sidney, and thus make him one of the most powerful nobles in England.[13]

At the Privy Council, Leicester led a Puritan group which included his brother, his father-in-law (the Earl of Bedford), Sir Francis Knollys, and two key Crown administrators: Sir Walter Mildmay and Sir Francis Walsingham (the Chancellor of the Exchequer and Principal Secretary of State, respectively). They urged upon the queen a forthright programme. In domestic affairs, they desired the weakening of the episcopal administration of the Church and royal acquiescence to unbeneficed preachers, lay preaching, and alternative liturgies, prayer books and Bibles. In foreign affairs, they pressed for the maintenance of an army in the Low Countries to defend co-religionists from the Spanish enemy, and the active, explicit, support of all Protestant states against Spain and the papacy. Yet to the annoyance of Leicester and his allies, the queen equivocated or (when pressed to a decision) chose most often to side with the cautious counsels of William Cecil, Lord Burghley.

In the late summer of 1576, after less than a year at Gray's Inn, Francis Bacon was placed under the tutelage of Sir Amias Paulet, who had just been appointed as the English ambassador to the French king. Paulet was close to Leicester's circle of councillors and courtiers rather than a political friend of the Lord Keeper but Paulet's embassy was a splendid chance for Sir Nicholas to provide Francis with a privileged education in the mysteries of royal service at a foreign court.[14]

Sir Amias' mission was a delicate one. He needed to monitor political activities in France, to mollify the French about the continued detention in England of Mary, Queen of Scots (who once had been a queen of France), and to maintain Elizabeth's discreet communications with the leaders of the Huguenot cause. In contrast with England, France in the 1570s was plagued by ambitious regional magnates, a weak central authority and by wars of religion. In some respects this state of affairs suited England very well, but a strong possibility now loomed that the ultra-Catholic faction (led by the cardinal and the Duke of Guise) might not only destroy the Huguenot king of Navarre but overwhelm the French royal government as well, forcing an aggressive foreign policy upon the young Valois king, Henri III. Most of the ambassador's stay was in Paris, for the obvious reason that the king spent much of his time there. An ambassador was expected to be in constant attendance upon the court, and we may assume that young Bacon as well as Sir Amias spent many hours mixing with the courtiers and officials of the Valois court.[15] Perhaps Francis attended upon Paulet during his formal audiences with the French king and his advisors; certainly, his close familiarity with Henri III's courtiers would make him aware of the king's philosophical *académie*.

Very recently, a group of noble courtiers and scholars had begun to meet with Henri III in order to hold debates about moral philosophy, poetry and music. As Dr Valentine Dale, Paulet's predecessor as ambassador, remarked disapprovingly to Secretary Walsingham, 'for all these troubles, the King has used of late to call certain poets and philosophers unto his chamber to hear them dispute three or four hours together *de primis causis de sensu et sensibili* and such like questions'.[16] This group called itself the Académie du palais, since such a name celebrated not only the involvement of the king himself but emphasised the kinship with other French *académies* (such as the Académie de poesie et musique and the Académie pléiade) and recalled as well the original Academy of Plato in ancient Athens.[17]

A principal concern of the scholars and courtiers in the French academies was the problem of how to attain states of virtue: the central aim of classical moral philosophy. Their neo-Platonist convictions led them to believe that 'true learning' in the nation's political leaders would be the best way to recreate a civil harmony in chaotic France and, eventually, to restore Christendom to its lost unity. In the Académie du palais of Henri III and his friends, the discussions of the means of personal moral improvement and religious fulfilment often turned toward associated issues of public ethics. They believed that the achievement of a *personal* harmony (through careful training of the natural capacities and through self-knowledge) would lead to the achievement of harmony *in the state*, through good government by the rulers and through the establishment of justice and order by the civil magistrates.[18] An emphasis of this sort was perfectly appropriate in a French Académie du palais, yet one should notice the great difference between the academicians' beliefs about how to achieve harmony in the state and those of Marsiglio, the Florentine Rucellai Gardens circles, Cromwell and his successors.

Francis Bacon remained at the French royal court for nearly two and a half years, but in February of 1579, just after he turned eighteen, news came to him that his father had died. Francis returned to England for the Lord Keeper's funeral on 9 March and he also returned to his rooms and studies at Gray's Inn, knowing he now would need to recommend himself to a patron in order to establish a career.

By 1579, Sir Philip Sidney and his close friends were meeting at Leicester House for discussions about politics, religion, moral philosophy and literature. They called themselves the 'Areopagus', which reflected their conscious imitation of Continental academicians' reverence for classical models and also indicated the purpose of their group. The 'Areopagus' had been the supreme judicial assembly of the ancient Athenians, and Sidney and his friends wished to debate and to legislate upon 'great points of law, God, or moral good'.[19] Soon after his return to Gray's Inn, Francis Bacon reacquainted himself with Leicester House. He had spent two and a half years at the French court; he had recent gossip about the king's *académie*; and he had met Sidney's old friends, the Huguenot poet-courtiers of Navarre, as well. Bacon had good reason to expect Sidney and the 'Areopagus' would have welcomed his news, and he had another, pressing, reason for visiting Leicester House: to ask for the patronage of the Puritan lords. His mother, Lady Anne, had thrown herself wholeheartedly into 'Godly' causes after her husband's death and she encouraged Francis to look to the Puritan lords for his political advancement.[20]

1579 and 1580 were anxious years for Leicester and the other English Privy Councillors. Not only was the queen engaged in the courtship of the Duc d'Alençon, the French king's brother (was it, or was it not, solely a diplomatic gambit?), but Catholic mercenaries landed in Ireland to stir up a revolt against the English and, in September of 1580, this band was reinforced by some Spanish troops. In 1580 as well, Spain overran the kingdom of Portugal, acquiring its deep-sea navy and its well-equipped ports, and Jesuits began to seep into England as 'missionaries' to the beleaguered English Catholics. During the winter of 1580, Queen Elizabeth decided to recall her parliament for a vote of supply and to pass stiffened laws against Catholics and especially against their priests, whom the government viewed (reasonably enough) as insurrectionaries.[21]

Leicester and the other Puritan leaders mustered their friends and clients for the forthcoming parliamentary session. This parliament had first met in 1572 and its last meeting had been in 1576; as many members had died or retired since then, by-elections were needed to fill their seats. In January of 1581, when parliament met again, two of its new members were Sir Philip Sidney and Francis Bacon, both of whom sat in the Leicester House interest. Bacon sat for Bossiney, a borough controlled by the Earl of Bedford. With no evidence of his activities in this parliament, we can only assume that Bacon fulfilled his patron's expectations, siding with the earl's clients and his Leicester House friends, and supporting the anti-Catholic legislative programme of the Puritan group. Lady Anne Bacon must have been gladdened by her son's political associations, but his uncle, Lord Burghley, may well have been displeased. The dying Sir Nicholas had entrusted his two youngest sons into Burghley's care: Lady Anne Bacon, it seems, disapproved enough of her powerful brother-in-law to ignore her husband's death-bed wishes. Francis was made an 'utter barrister' of Gray's Inn in June of 1582: a speedy advancement, considering he had been a student of the law for less than four years, and the 'call to the bar' of an Elizabethan Inn rarely came before seven years of residence.[22] It suggests that the benchers responded favourably to external persuasion on young Bacon's behalf. Given the rivalries between Burghley and the Leicester–Walsingham group of councillors, there is little reason to suppose that Bacon's advancement was due to the efforts of Lord Burghley rather than those of his patrons in Leicester House.[23]

In this period (1581–3), Sir Philip Sidney was writing a justification of his 'unelected vocation', poetry, in which he argued not only that poetry was an honourable art but that it was uniquely valuable among all the arts and sciences:

> This purifying of wit, this enriching of memory, enabling of judgement, and enlarging of conceit, which we commonly call learning, under what name soever it come forth, or what immediate end soever it be directed, the final end is to lead and draw us to as high a perfection as our degenerate souls, made worse by their clayey lodgings, can be capable of ... So that, the ending end of all earthly learning being virtuous action, those skills, that most serve to bring forth that, have a most just title to be princes over the rest.[24]

Sidney's manuscript, known as 'The Defence of Poesy', circulated among his friends for their comments and served as a basis for their philosophical arguments. The 'Defence' can thus be taken as further illustration of the sorts of discussion which were taking place in Sidney's 'Areopagus' circle when Francis Bacon was associated with Leicester House in the early 1580s. These discussions concerned not only 'law, God, and moral good' (as Daniel Rogers had reported) but also the nature and purposes of human learning and knowledge.

In late 1584, the queen called for a new parliament. A pointed indication of Bacon's political allegiances was made before the parliament met in November: Burghley offered Bacon a seat

(Gatton, in Surrey), and he was returned by the dutiful voters there, but he refused to accept his uncle's patronage, and once again sat for a borough in Bedford's control. Francis Bacon had again displayed himself a loyal member of the Leicester House Puritan grouping.[25]

From the time the election writs went out in the autumn until the closing of the parliament at the end of March 1585, the Puritan leaders in the Lords and Commons waged 'a political campaign without precedent in parliamentary history'.[26] This parliament was dominated both by fears that Catholic zealots might murder the queen and also by 'Godly' anger at the authoritarian programme of the new Archbishop of Canterbury. Protestant suspicions of English Catholics had been heightened in recent years by a continuing infiltration into England of Jesuits and young seminary priests. These suspicions grew into near-hysteria after the government revealed in June 1584 the details of the Throckmorton plot to kill Elizabeth, and with the news in July that a Catholic fanatic had succeeded in the murder of William of Orange, the famous Protestant prince in the Netherlands.

This alone was alarming enough but, in addition, it had been none other than Dr John Whitgift who had been elevated to the Archbishopric of Canterbury in August 1583. He swiftly moved to assert episcopal discipline within the Church, to suppress the Puritan clergy's training 'conferences', and to stop the activities of non-beneficed Puritan preachers and lecturers. 'Godly' clergy and laymen, badly shaken by the vigour of Whitgift's assaults, were determined to use the new parliament for a counter-attack on what they saw as a reactionary policy for the Church. They recognised that new legislation alone could stop Whitgift's plans to enforce all the provisions of the religious settlement of 1559. They also recognised that the archbishop must be curbed quickly, since it was apparent that his attacks were causing their movement to dissolve into moderate and extreme factions.

As part of this excited political campaign, several MPs wrote treatises hoping to sway the queen away from full support of the archbishop, and among these petitioners was none other than Whitgift's former pupil, Francis Bacon.[27] Bacon's presumptuous manuscript was entitled 'A Letter of Advice to the Queen', and the fact that several copies survive suggests his 'Letter' had a fair circulation.[28] The 'Letter' had two principal themes: that English Catholics were dangerous and their numbers and strength should be lessened, and that there was a need for an international league against Spain, followed by a war against her. Bacon began by mentioning 'these late wicked and barbarous attempts', presumably referring to the Throckmorton and Parry plots against the queen. He claimed of the queen's safety that 'my life shall be among the foremost to defend it', a claim which suggested that he, too, had taken the 'Oath of Association' that winter,[29] and he went on to echo Sir Walter Mildmay's speech of 12 March about the need to avoid overly savage anti-Catholic legislation. Severe penalties, declared Bacon, would prove counter-productive: 'though [Catholics] must be discontented, yet I would not have them desperate: for among many desperate men, it is like some one will bring forth a desperate attempt'.[30]

Like all the reforming Puritans in parliament and Convocation, Bacon desired the maintenance of 'careful and diligent preachers in each parish', and he criticised Whitgift's drive to root them out:

> In their opinions though they are somewhat over-squeamish and nice, yet with their careful catechizing and diligent preaching they bring forth that fruit which your most excellent Majesty is to wish and desire; namely, the lessening and diminishing of the Papistical number.

Whitgift recently had asserted in the Lords that Puritan preachers would eventually 'bring the Church to an anarchy';[31] but this argument was discounted by Bacon, who remarked that 'for those objections, what they would do when they gat a full and entire authority in the Church, methinks they are *inter remota et incerta mala*'.

As for his second major theme, foreign strategy, Bacon's opinions were very similar to those attributed to Sir Philip Sidney. Sidney apparently believed the Italian states would rise up against their Spanish masters if alliances were offered to them.[32] In the 'Letter of Advice', Bacon claimed that England's strength 'must be in joining in good confederacy, or at least intelligence, with those that would willingly embrace the same', and he referred not only to Protestant states but to Catholic, Italian ones ('Florence, Ferrara, and especially Venice') which 'fear and abhor the King of Spain's greatness'. Just like Sidney, Bacon desired attacks on the king of Spain's outlying possessions, 'thoroughly and manifestly both upon his Indies and Low Countries, which would give themselves to you'.[33]

The young Francis Bacon can be identified, then, with the principal men and the policies of the Puritan movement. He had accepted the parliamentary patronage of the Earl of Bedford – not once but twice – and in doing so he rebuffed similar offers from his powerful conservative uncle. He became associated with Leicester's celebrated nephew, Sir Philip Sidney, and worked with Sidney in fostering connections with Henri of Navarre's courtier-poets.[34] His 'Letter of Advice to the Queen' was part of the great Puritan campaign of the 1584–5 parliament, and in it Bacon echoed the opinions of the political leaders of the 'Godly', who believed the battle for a 'Godly Church' and for the destruction of the 'papists' was a principal part of their duty as statesmen.

Yet the queen rejected the parliamentary petitions about religion and, despite the misgivings of many in her Privy Council, she offered her archbishop her complete support. When Whitgift's bold policy of confronting 'Godly' reformers brought protests from her Puritan councillors, and this provoked a declaration from her, Elizabeth made it abundantly clear that she detested Puritans and that she now was prepared to resist alterations in her established Church. Here began the rapid disintegration of what had long been a broad alliance of reforming Protestants in Elizabethan England. Opposed demands, from the queen and archbishop (insisting on a strict conformity) and from the zealous (insisting on greater 'purity' in worship and Church structure), soon undermined any general consensus among the reform-minded by exposing the wide diversity of goals held by individual English Protestants. This was a crisis for men and women who desired religious reforms: could they justify disobedience to the expressed wishes of the Supreme Governor of the Church? Most said no and submitted, but some remained obdurate. Disagreements among those in the reformist ranks, once easily overlooked as minor, now became notorious and divisive. Unbowed by the queen or by the erosion of support, Leicester and his staunch allies plunged ahead: fervent Puritan clergymen began secretly to organise themselves in Presbyterian fashion, and their leaders became both more vocal in their demands and more self-consciously an anti-episcopal 'opposition' faction.[35] As Whitgift's grip on his archbishopric grew, Leicester's earlier power of ecclesiastical patronage declined. Making virtue of necessity, Leicester began to criticise the continued existence of ecclesiastical benefices. Commoners like Mildmay and Walsingham, whose political powers were fully dependent on their royal administrative offices, soon disassociated themselves from active support of the movement. For Walsingham at least, anxieties about the internal security of the state began to take pride of place.[36]

This crisis and fragmentation in the Puritan movement brought a crisis for the young and ambitious Francis Bacon; he now was faced with a grave career decision. It had become clear that his noble patrons (and his mother as well) were setting themselves against the queen's explicit wishes. Bacon could no longer expect their patronage to gain him office in government and certainly, when compared with the broad 'commonweal' aspirations he had learned from his father, their ambition now seemed narrowly sectarian.

With the exhaustion of Elizabeth's diplomatic efforts concerning the Low Countries, the antagonisms that had long existed between England and Spain finally became open war. Leicester was given his long-held wish: he was to lead an English force against the Spanish enemy. In the autumn of 1585, Sidney, and then Leicester and the army sailed over to the Continent.[37] While his patrons and courtly friends campaigned in the mud of the Netherlands, Bacon remained behind, observing rapid changes in the political scene at court. During the coming year he abandoned his associations with Leicester House and with a newly radicalised Puritan movement.

A conservative reaction swept through the places of power in 1586. While the fighting sputtered fitfully in the winter months, and while the government prepared for a Spanish invasion, the character of the Privy Council was altered drastically. In the absence of the domineering Leicester, the queen swore in three conservative members: Lord Cobham (her cousin), Lord Buckhurst (an ally of Burghley's) and Archbishop Whitgift. These additions powerfully reinforced the influence in the council of conservatives such as Sir Christopher Hatton. A month later, the immensely popular Puritan preacher, Walter Travers, was suspended from his preaching duties at the Middle Temple in favour of the conformist Richard Hooker.[38] It was absolutely clear that the political winds in the queen's court and council were blowing strongly against the 'Godly'. On 21 September 1586, Sir Philip Sidney died of festering battle wounds. His death deeply shocked the political nation. Grief and mourning were widespread, but the death of this charismatic leader was an especially cruel blow to the hopes of his noble uncles and the radical religious reformers. Only ten days later, the queen furthered the conservative tilt of her Privy Council by advancing Dr John Wolley, her Latin secretary and a civil lawyer.[39] Puritan political influence at court had been damaged very severely.

When writs went out in October 1586 for elections to a new parliament, Bacon declined the Earl of Warwick's offer of a Commons seat in the Leicester House interest. He accepted instead a seat provided by the Bishop of Winchester, Thomas Cooper, a respected scholar, administrator and a friend of Mildmay and Walsingham.[40] Despite the obvious favour now given by the queen to conservative, 'conformist' politicians, Bacon did not make a cynical *volte face*. He retreated from what was now a radical faction and sought instead the patronage of senior figures whose overriding characteristic remained unswerving dedication to the business of governing the queen's realm.[41]

In statesmen's eyes, England was facing now a desperate crisis which took precedence over any other issue. Despite the subtle diplomacy of past decades, England was at war with Christendom's wealthiest and most powerful state. The duty of Elizabeth's councillors and loyal servants was clear: to put her people in a state of preparedness for war, to coordinate money and resources for battle and to defend the realm against enemy attack. It was now that Bacon found his first official employment: the investigation of English Catholics. Through the patronage of Sir Francis Walsingham, Bacon began several years' service in gathering domestic intelligence about Jesuits, seminary priests, their accomplices and their plots to foment recusant uprisings. Walsingham was the master of the government's intelligence, 'a most subtle searcher of hidden secrets'.[42] Bacon's legal training made him a valuable assistant to Walsingham in this sort of work, since much of it appears to have consisted in disentangling fragmentary or contradictory information from a web of informers, in preparing 'schedules of interrogatories' for witnesses and suspected Catholic plotters, and in presenting reports of their examinations to his superiors.[43]

It was immediately clear that Bacon not only exerted himself greatly on behalf of his new patrons but that he was deemed by them entirely trustworthy. During the parliament of the winter of 1586–7, which concerned itself principally with the fate of Mary, Queen of Scots, Bacon was visible for the first time as a member of government. For example, he was among the first speakers both in advocating Mary's execution and in urging the proposed subsidy; that he was allowed a senior place in the queue to speak on the two essential matters of the parliament is an indication of his changed status in the Commons. Judging by the earliest surviving letters to Bacon from the Privy Council, Bacon's efforts in legal investigations as well as in parliament inspired great confidence. In one such letter, 'their Lordships' desired the legal opinions of the Attorney-General, the Solicitor-General, and 'Mr Francis Bacon, Esq.', upon various enclosed examination reports of two Catholic prisoners.[44] His enhanced standing was also manifested by his rise within Gray's Inn during 1587. In February of this year, he was permitted the dining privileges of the readers although he remained an utter barrister. Bacon's anomalous status was quickly removed: he was soon made a bencher of Gray's and, in November, he was elected as the next reader.[45] In August 1588, as the broken Armada was scattered in the North Sea, Bacon and eight other men (including three High Commission members) were ordered to examine *all* the recusants then in gaols in London, to pass judgement upon their cases and, furthermore, to establish if any were Jesuits or priests.[46] Bacon now was regularly involved in this grim and profoundly serious judicial work.

Bacon soon received more signs of conciliar confidence in his legal and administrative abilities. In late 1588, the councillors requested advice about repealing or reforming 'all such Statutes as are found unnecessary or defective'. Letters went out to the Chief Justices of the King's Bench and the Common Pleas, to the Attorney-General and Solicitor-General, and to four men from each of the four Inns of Court. Francis Bacon was one of this group of twenty; he was firmly among the Crown's trusted legal advisors.[47] In early 1589, when yet another parliament was called, Bacon took a seat (Liverpool) in the gift of the Duchy of Lancaster. The duchy usually attempted to seat its own senior officials and government friends, and this election was no exception; moreover, the new chancellor of the duchy was Bacon's patron, Sir Francis Walsingham.[48] In August, the Huguenot Henri of Navarre, whom Elizabeth had long supported against the Catholic Valois monarchs and Philip II of Spain, was crowned king of France. Shortly after this, Bacon prepared on Walsingham's behalf a statement for the benefit of the new French court which would justify the queen's religious policies, particularly her recent assault against the Puritan reform movement. The Huguenot leadership sympathised strongly with the English Puritans, and the Privy Council's response would be a matter of considerable diplomatic delicacy. That the council at this point in time should have commissioned Bacon to write a formal account of the queen's policies illustrates yet again both his versatility and the high regard in which he was held.[49]

In the autumn of 1589, perhaps by way of recognition for his labours, Bacon received the reversion to 'the Clerkship of the Council of State', otherwise known as the 'Clerkship of the Privy Council in Star Chamber'.[50] The Court of Star Chamber was concerned with enforcing the royal peace and royal justice. It was, in fact, the central government's principal instrument for impressing obedience to the strict rule of law upon the realm and for punishing all outbreaks of 'riot and affray'. The Clerk of the Star Chamber was the most senior of the various clerks who surrounded the councillors, and he was the only one who needed considerable legal (as opposed to diplomatic and administrative) experience. 'It was the duty of the clerk to receive, endorse, enter, keep, and certify the bills, pleadings, records, orders, rules, sentences, and decrees of the Court', but much of this work was carried out by subordinate officers.[51] In practice, the clerk functioned as a judge of process. The councillors (with the Chief Justices) sat in the Star Chamber on Wednesday and Friday mornings to hear attorneys argue cases and to pass down judgements; the clerk, however, 'sat alone to see ordinary course observed in prosecution of causes upon all the days of the week but when the lords sat to hear causes'.[52] Bacon's reversion held promise of great responsibility in the judicial apparatus of the state. When it 'fell in', not only would he become a minor judge and a close associate of the principal statemen of the realm, but the legal problems he would deal with were explicitly those concerned with the security and strength of the state.

In 1589 a radical Puritan satirist and polemicist known as 'Martin Marprelate' captured the attention of the reading public and provoked a huge row about the legitimacy of the Church's government.[53] Marprelate caused severe anger among his ecclesiastical victims, not only for his witty attacks upon their callings and persons but also for his ribald handling of religious issues. Privy Councillors were concerned for additional reasons. Not only had Marprelate secreted a printing press (itself a crime) but the councillors, ever vigilant against possible rebellion, viewed the widespread popular amusement with which the tracts were greeted as ominous indirect criticisms of their regime. Any hint of rebellion or riot was always a cause for alarm but now especially so, when the government's human and material resources were fully engaged in the war efforts.[54]

Bacon wrote an 'Advertisement' concerning Church affairs during the latter part of the year. We can see in it both his stout committment to moderate, as opposed to Presbyterian or more radical, religious reform and also his support for the Church hierarchy and government. Such convictions might seem to rest uneasily together but the 'Advertisement' suggests that Bacon's view of the controversies and of the purpose of Church government was essentially a secular one. His arguments imply that the Church government was legitimate because it was a *part* of the royal machinery for governance. He defended the members of the hierarchy for their learnedness and their (royal) office, for both of which they deserved respect and obedience, and he defended their ministrations in so far as they appeared to support the peace and prosperity of the realm. Holding to this pragmatic and secular point of view, Bacon was able both to support the Church *qua* government against radical critics and to criticise certain present policies of the Church as well.

The public attacks and counter-attacks of radicals and conformists had been extremely bitter. Bacon was deeply alarmed but he was not interested in the 'rightness' of rival arguments; his paramount concern was the continued peace and harmony of the community, and he viewed the controversy principally from a secular statesman's standpoint.

Judging from the careful structure and polished style of 'An Advertisement Touching the Controversies of the Church of

England', Bacon intended to circulate it.[55] A striking feature of this treatise is the elevated perspective Bacon took for his survey of the Church's current troubles; he is not an advocate here but a statesman and, accordingly, 'the controversies themselves I will not enter into, as judging that the disease requireth rather rest than any other cure'.[56] Instead, he proceeded to summarise what he believed to be the general causes of the present conflicts and the general faults of the parties concerned. His analysis, again, was a secular one: 'we contend about ceremonies and things indifferent; about the extern policy and government of the church'. These matters of controversy, then, properly lay within the domain and competence of the secular statesman. Moreover, Bacon suggested he was a judge, proceeding by equity to resolve stubborn conflict:

> I find in reason, that peace is best built upon a repetition of wrongs ... So it is true which is said, *Qui pacem tractat non repetitis conditionibus dissidii, is magis animos hominum dulcedine pacis fallit, quam aequitate componit.*[57]

Perhaps it was solely his three years' employment in examining Jesuits and other Catholics which had suggested this might be the only way to defuse religious zealotry, yet it seems likely Bacon also remembered the example of his father, who as Lord Keeper, the Crown's senior judge, regularly proceeded by equity rather than strict common law, precisely so as to conclude otherwise intractable disputes.

Bacon stressed the moral duty to preserve the peace of the community; he advised 'every man be swift to hear, slow to speak, slow to wrath' and he warned that the controversies were undoing 'the bonds of peace'. He then demanded an end to the propaganda battle ('this immodest and deformed manner of writing lately entertained, whereby matters of religion are handled in the manner of the stage') and roundly scolded both 'Martin' and Dr Bancroft, Whitgift's assistant:

> To turn religion into a comedy or satire; to search and rip up wounds with a laughing countenance; to intermix Scripture and scurrility sometime in one sentence; is a thing far from the devout reverence of a Christian, and scant beseeming the honest regard of a sober man ... I dislike the invention of him who (as it seemeth) pleased himself

> in it as in no mean policy, that these men are to be dealt withal at their own weapons, and pledged in their own cup.[58]

Bacon affirmed his respect for the Church hierarchy. He supported the bishops ('no contradiction hath supplanted in me the reverence I owe to their calling; neither hath any detraction or calumny embased mine opinion of their persons'), and he explicitly condemned the Presbyterian ideal ('the parity and equality of ministers is a thing of wonderful great confusion; and so is an ordinary government by synods'). He declared, furthermore:

> Our church is not now to plant; it is settled and established. It may be, in civil states, a republic is a better policy than a kingdom: yet God forbid that lawful kingdoms should be tied to innovate and make alteration.

He asserted hereby not only that the legitimacy of the present Church government rested on secular legal grounds (that is, it had been *settled* upon the queen and *established* by the parliamentary Act of Supremacy in 1559), but he again displayed his politician's perspective upon these religious controversies: in the radicals' demands upon the Church, Bacon saw rebellious and republican implications *for the state*.

His conservative anxieties and justifications for the regime notwithstanding, Bacon was convinced that the bishops and the Church had much to answer for:

> To my lords the bishops I say, that it is hard for them to avoid blame (in the opinion of an indifferent person) in standing so precisely upon altering nothing ... Is nothing amiss? Can a man defend the use of excommunication as a base process to lackey up and down for fees and duties; it being the greatest judgement next the general judgement of the latter day? Is there no means to train up and nurse ministers (for the yield of the universities will not serve, though they were never so well governed) – to train them, I say, not to preach (for that every man confidently adventureth to do), but to preach soundly, and handle the Scripture with wisdom and judgement?

Like most critics of the Church, Bacon complained about the retrograde confusion of revenue and spiritual discipline and about the failure to recruit enough clergymen to discharge the Church's pastoral duties. His criticisms were sharp, yet they remained criticisms of 'extern policy and government', of Whitgift's refusal to concede any popular alterations in the institution. Certainly, 'sound preaching' was the central desire of all moderate reformers; like the more radical Puritans, they wished to see more preaching of the Word, but they held particular ideas about doing it 'soundly'. Bacon's view of an effective minister was a man who would impart to his flock not 'matters of controversy' but positive directions for daily Christian living:

> It is an easy and compendious thing to call for the observation of the Sabbath-day, or to speak against unlawful gain; but what actions and works may be done upon the Sabbath, and in what cases; and what courses of gain are lawful, and what not; to set this down, and to clear the whole matter with good distinctions and decisions, is a matter of great knowledge and labour.

Bacon's minister, dispensing 'doctrine of manners', as he termed it, and providing 'distinctions and decisions', would be engaged in a quasi-judicial activity, no different in purpose than the work of his secular cousin, the Justice of the Peace, the other crucial officer of Elizabethan local governance.

As could be expected, Bacon's picture of the radical Puritans was a highly critical one. He portrayed them as self-righteous and stubbornly disrespectful:

> They have impropered to themselves the names of zealous, sincere, and reformed; as if all others were cold, minglers of holy things and profane, and friends of abuses ... And as they censure virtuous men by the names of civil and moral, so do they censure men truly godly and wise (who see into the vanity of their assertions) by the names of politiques; saying that their wisdom is but carnal and savouring of man's brain.

Bacon's elitist social beliefs extended to learning as well as to governance. In 'An Advertisement', he declared 'the people is no meet judge nor arbitrator, but rather the quiet, moderate, and private assemblies and conferences of the learned'. In a short tract written at this time, he condemned those who came to an opinion without sufficient evidence and an experience of affairs:

> Let none that is unlearned presume to admonish another in controversies of religion. Let not one that liveth alone and seeth not

> into the affairs of the world presume to advise others of their proceedings.[59]

Knowledge and authoritative judgements, then, were possessed only by those few who had great learning.

These general beliefs Bacon shared with most members of the governing classes, yet some of his more specific convictions on the subject of learning were novel and he never abandoned them. In particular, there was in Bacon's criticism of the radical Puritans also a social and psychological explanation of the formation of sects and intellectual prejudices in general. Certain vain men – lovers of 'pre-eminence' – seek 'inward authority ... over men's minds, in drawing them to depend upon their opinion, and to seek knowledge at their lips'. Such men attract 'stiff followers', usually young men of 'superficial understanding, carried away with partial respect of persons, or with the enticing appearance of goodly names and pretences'. Around these 'general affections' were wreathed 'accidental and private emulations and discontentments' and Bacon's bleak conclusion – drawn from his experiences both when Whitgift's Cambridge pupil and from his present security employments – was that 'few follow the things themselves, more the names of the things, and most the names of their masters'.[60] In elaborating these points, he declared that most men

> leap from ignorance to a prejudicate opinion, and never take a sound judgement in their way ... when men are indifferent, and not partial, then their judgement is weak and unripe through want of years; and when it groweth to strength and ripeness, by that time it is forestalled with such a number of prejudicate opinions, as it is made unprofitable: so as between these two all truth is corrupted.

'General affectations' and 'accidental emulations and discontentments', Bacon believed, 'together break forth into contentions, such as either violate truth, sobriety, or peace.' Since these resulted in 'prejudicate opinions', as well, we reasonably may take Bacon to be implying that 'contentions' are manifestations of 'prejudicate opinions'. Moreover, since 'prejudicate opinions' forestall sound judgements, men's determinations are usually 'unprofitable' and 'unripe'; that is, they neither proceed from, nor do they provide, knowledge of 'truth', 'sobriety' or 'peace'.

Bacon's argument has a crucial further implication: since 'prejudicate opinions' (and 'contentions') violate not only 'truth' but also 'sobriety' and 'peace' (which were for Bacon, given his security work, *political* values), then not only were learning and knowledge fraught with problems but the nature, causes (and errors) of learning and knowledge were properly among the concerns of a conscientious statesman, and especially proper for a prospective Clerk of Star Chamber.

Not only had Sir Nicholas prepared Francis Bacon for the law and for royal service but he had impressed upon him what he himself believed: the essential duties of a royal servant were to provide the monarch with good counsel and to advance the commonweal of the realm. In addition, Sir Nicholas' judicial career and concerns taught his youngest son that the rule of law and the even-handed dispensation of justice were crucial for the peace and prosperity of the kingdom.

Lady Anne Bacon, devoted to Puritan causes, encouraged Francis to seek his career through association with Puritan noblemen and to distance himself from Sir Nicholas' old colleagues who, she thought, had grown lukewarm in their encouragement of 'Godly' reforms. But when the queen rejected Puritan attempts to pressure her and the Puritan leaders became increasingly radical, Bacon (then twenty-five years old) felt forced to made a serious choice. Unsurprisingly, he chose his father's career, his father's political values, and the patronage of councillors who resembled his father both in their dedication to the queen's service and in their cautious reforming instincts.

Nor was it surprising that Bacon should have come to believe the problems of knowledge were within the brief of a statesman-judge. Everywhere he turned, he found his elders discussing knowledge and the relation of learning to virtuous public action. When at Cambridge, his tutor had argued that wise men were guided by 'sure proofs and certain knowledge' and not by 'affectation, opinion, and fancy'. The members of the French Académie du palais believed there were direct parallels between the search for self-knowledge and an eventual personal virtue and harmony (on the one hand) and the search for vigorous royal governance and the rule of justice and order in the state (on the other). Sir Philip Sidney, whom Bacon knew well in London, had agreed with the Continental academicians and argued that it was 'poesy' that best served 'to lead and draw us to as high a perfection as our degenerate souls are capable of'. Despite particular misgivings about his former tutor's policies for the Church, Bacon did come to accept Whitgift's beliefs that radical religious faction was corrosive of the unity of the community and that such faction began in individual self-delusions and pride. His immersion in the problems of security for the realm (and, specifically, in the interrogation of religious zealots) led him to view 'contentions' as manifestations not solely of intellectual 'dis-orders' but of *political* disorder. Similarly, his increasing self-identification as a legal counsellor led the Lord Keeper's son to value 'sure proofs and certain knowledge' more than 'opinions', nearly all of which he deemed, in his revealing term, as 'prejudicate'.

By 1590, then, Bacon was not only engaged in royal service but anticipating major legal and judicial appointments. He was given to statesmanlike concerns about learning and knowledge. He had identified certain general causes and public effects of the problems of learning. He believed with Sidney that 'the ending end of all earthly knowledge [was] virtuous action' although, unlike Sidney, he was as yet uncertain which 'skills' would best serve to bring it forth. But Bacon's career plans were thrown into disarray in the spring of 1590: on the sixth of April, Walsingham, his recent patron, died.

3

Business of state

> I houlde every man a debter to his profession, from the which as men of course doe seke to receave contentments and profit, so ought they of duty to endeavor themselves by waye of retribution to be a helpe and ornament therunto.
>
> *Maxims of the Law* (1596), C.U.L. MS Ff.4.16, fo. 7

Having come by 1590 to the conviction that problems of knowledge were a proper part of a statesman's concerns, Bacon soon made substantial elaborations upon this general principle. In 1592 and 1595, he argued not just for the consideration by statesmen of knowledge *per se* but specifically that a 'properly' reformed and organised natural philosophy could transform the commonweal of England.

It would be pleasant indeed to know precisely when it was that Bacon first came to the particular convictions he expressed at this time, but such specific information is beyond our recovery. Nonetheless, we can recover considerable information about Bacon's activities in the early 1590s, and it is sufficient to answer a related and pressing question. Why was it that Bacon chose to make these declarations about natural philosophy when he did? In other words, what local concerns prompted him to write? Since amid the documents of these years are Bacon's earliest surviving statements about natural philosophy, we need to address these queries if we are to understand what Bacon was saying. Because these writings were *Bacon's* responses to *his* circumstances, we must consider the situation in which he found himself in those years before turning to examine his pronouncements. By proceeding in this way, we find that Bacon's early writings about natural philosophy display his reaction to developments he regarded as *politically* dangerous and a serious challenge both to the stability and to the 'commonweal' of the realm. In Bacon's eyes, natural philosophy could be a solution to problems facing statesmen: it could be refashioned into a splendid support for the Tudor state.

In April of 1590, Bacon's future as a servant of the Crown was thrown into doubt by Walsingham's death. It was true that Bacon now was known to be a reliable advisor in recusancy examinations and prosecutions, but however much the councillors worried about Jesuits and Catholic subversives, Bacon's employment remained irregular, at the periphery of the court, and dependent upon the continued good will of councillors. He did hold a major reversion – to the office of Clerk of Star Chamber – but he could not expect that it would conveniently fall in soon. In short, Bacon was not yet launched into a career as a statesman. He lacked any office by which he himself might construct his further rise at court, and he remained in need of help from a sympathetic patron.

Fortunately for Bacon (or so it seemed), such a person existed: Robert Devereux, the young Earl of Essex. In Bacon's eyes, Essex was not merely one among several possible patrons at court, nor was he to be sought after simply because of his well-known generosity. In fact, Bacon saw Essex as his obvious patron. Essex was widely regarded at the beginning of the 1590s as the rightful successor not only to the Earl of Leicester, but also to Sir Philip Sidney *and* to Secretary Walsingham. Poor Essex quickly came to carry the great burden of other people's expectations, including those of Francis Bacon, who later declared that 'I held at that time my lord Essex to be the fittest instrument to do good to the State'.[1]

Yet why was Essex expected to adopt the mantles of Leicester, Sidney and Walsingham? It was true that celebrated marriages had linked the four men: Leicester had married Essex's widowed mother, and Sidney's widow (Walsingham's daughter) became the young Earl of Essex's bride, but this in itself hardly was compelling. Part of the reason must lie in Essex's own ambitions for himself and his country, nurtured by his high birth, by his intimate association for several years with both Leicester and Sir Philip Sidney, and by the favour showered upon him at court by the aged queen. But equally (if not more) important were the political pressures which built up as the old leaders of the Protestant 'war party' died, even as their most enthusiastic and vunerable supporters were under attack from Whitgift. The Earl of Bedford had died in 1585, Sir Philip Sidney a year later; in early 1588 died John Field (the radical

Puritans' organising 'secretary'), and then – during the eighteen months following the defeat of the Invincible Armada – Leicester, Sir Amias Paulet, Sir Walter Mildmay, the Earl of Warwick and Secretary Walsingham.[2] Burghley, no longer jostling with them for favours and power, sought to gather up the patronage power of his dead colleagues and to draw ever closer around him the conservative newer members of the council. Always suspicious of overmighty councillors, Elizabeth previously had played them against each other but now, with few of her long-trusted servants still alive, a government dominated by Burghley seemed increasingly likely. It was thus as the standard-bearer of the court's Puritan 'war party' and as a brake against Burghley's pre-eminence that Essex swiftly became the object of many hopes.

Essex had gone in late 1585 with his step-father, Leicester, to fight the Spanish in the Low Countries. He remained there for nearly two years, becoming skilled and admired as a military captain and becoming increasingly enamoured with the soldiering life. In September of 1586, he won his spurs at Zutphen, fighting boldly beside Sir Philip Sidney. Because Sidney died of wounds received at Zutphen, gentlemen who fought there with him shone in his reflected glory. Yet for Essex, Sidney's friendship, and his death, meant so much more than this. A codicil to Sidney's will irrevocably established Essex's public fame and self-image as a Protestant champion: Sidney bequeathed Essex his best sword. No more powerful symbol of succession could have been devised: Essex had been anointed, as it were, by the dying hero himself.[3]

Soon after surrendering command of the armies in the Low Countries, Leicester had begun to surrender some of his position at court to Essex. In December of 1587, Essex had succeeded his step-father as Master of the Queen's Horse, an office (and a succession) which Elizabeth's courtiers knew was a highly significant one. Not only did the Master of the Horse have the right of personal attendance upon the queen but this had been the first office the young queen had given to Leicester at the beginning of her reign. Essex's special standing at court was underlined in April 1588 when he became a Knight of the Garter. It was then clear to everyone (except perhaps to Sir Walter Raleigh, the Captain of the Guard) that the dashing young champion was the queen's favourite, and that he was well placed to ask for (and receive) gifts and privileges for himself and others. Essex, like Leicester before him, was thus in a splendid position to accumulate political power.[4]

Because Leicester's, and Sidney's, promotion of Essex had been so clear and the queen's favour so public, Essex was seen by hard-pressed radical Puritans as the man to petition for aid and patronage. This long had been the position occupied by Leicester, and within a week of Leicester's death in September 1588, the aged Sir Francis Hastings prayed Essex would employ

> your whole forces bothe in honour and credite to advaunce [God's] truth and to be a comforte to his people professing the same. And now, my lord, is the time come wherin you shoulde put this in practise, in that He hathe taken from us that honorable worthy gentleman, whom He used many times as a notable instrument for the good bothe of the churche and commonwealth; and if it may please God to put into your lordship's hearte not only to succede him but to overgoe him in his care in this behalfe you shalbe most happy.[5]

Hasting's pleas that Essex should take up Leicester's burden were willingly received. There is ample evidence that Essex not only sympathised with the plight of radical Puritans but that in these years he did everything he could to assist them. By 1591, a courtier could write that 'both soldiers and puritans wholly rely upon him'.[6]

Essex also inherited something of Leicester's role as a patron of scholarly learning. Leicester had tried to pass to Essex the chancellorship of Oxford University. This was a bold and provocative move (Essex was hardly a venerable candidate), but Lord Burghley, Leicester's old rival, was chancellor of Cambridge, and it seems that Leicester planned to live beyond the grave through Essex, who therefore was made heir not solely to Leicester's place but to his rivalries as well. This attempt was unsuccessful (as was another made by Essex's friends in 1591), but ambitious scholars who previously had dedicated their books to Leicester began now to dedicate their works to Essex.[7]

With Leicester's death, it fell principally to Secretary Walsingham to urge the Puritan foreign policy at the Privy Council table (namely, close alliance with Protestant states and furtherance of the war against Spain), and Essex, the Protestant champion, became closer and closer allied with him.[8] In 1589, Essex's alliance with

Walsingham and his role as Sidney's successor was made even stronger when he secretly married Lady Frances Walsingham, Sidney's widow. This was an imprudent marriage; it was far below his station and without wealth – just as it had been for Sidney himself – but it signalled his deepening commitment to the roles fostered upon him by his step-father, by Sidney and by their supporters. Essex wanted desperately to do something for his country but he thought this was to be done in battle rather than by politicking and diplomacy. He was irritated increasingly by living 'idle' at Elizabeth's court, as his remarks to François De la Noue, one of Henri IV's foremost generals, serve to illustrate:

> I should be very happy to see some opportunity by which we could together win honour and serve the common weal. I am idle here and have nothing to do but to hearken for such opportunities.[9]

But with Walsingham's death, Essex's career, no less than that of Francis Bacon, was thrown into doubt. Essex was only twenty-three years of age and now he was at court without the protection and advice of a supportive Privy Councillor. He did have the favour of the queen – at the present – but it also brought him jealous enemies, and his known sympathy to the 'Godly' could quickly lead to isolation in an increasingly conservative court and then to the frustration of his ambitions. Moreover, with the death of those councillors who had supported him and whose aims he now championed alone, Essex needed something of their powers and authority if he was to proceed successfully. He needed to be more than a warrior: he needed, in fact, to become a Privy Councillor. It probably is here, in 1590, and in the dilemma facing the court's 'war party', that the fortunes of the Earl of Essex and those of Francis Bacon became enmeshed.

Despite the great favour shown him by the queen, Essex had little independent political importance: he had only a small territorial base – in Wales (and that was challenged by the Earl of Pembroke) – and little landed income.[10] A nobleman's membership of the Privy Council generally reflected his great independent power in the country and the need to placate or co-opt him. A commoner's inclusion usually reflected his senior position within the machinery of royal governance. Yet Essex possessed neither Leicester's independent power and resources nor Walsingham's diplomatic and administrative skills with which to bolster a claim to a place in the Privy Council. It was at some point during this year that Essex became persuaded that the best path to power was to emulate the old Secretary by controlling 'intelligence'. Walsingham's mastery of 'intelligence', whether domestic, foreign, 'political' or 'religious', had been crucial in his rise to a senior place in the Privy Council. The timely information produced by his private network of 'intelligencers' throughout western Europe had made Walsingham indispensable to the queen (however much his dour Puritanism might irritate her). By the autumn of 1590, Essex was using his own servants to collect 'foreign intelligence',[11] and his close friend, Sir Henry Unton, was appointed ambassador to France in July 1591, thereby becoming a splendid source of foreign news for Essex.

Walsingham's former clerks and intelligencers were eager for employment as well, and among these was Francis Bacon. Bacon and the Earl of Essex soon came to an agreement, for by April of 1591, Bacon was actively organising an intelligence network for Essex; that is to say, he was reactivating Walsingham's networks and re-employing many of his clerks and informants at home and abroad.[12] Arrangements were advanced sufficiently by July for Bacon to be again receiving spy information on priests and Jesuits.[13] It was natural for Bacon to turn to Essex. Bacon's political allegiances overlapped with those of the young Essex, and his employments under Walsingham had provided just the sort of skills that the earl – and Bacon – believed could eventually bring Essex political power. Bacon was as concerned now with security and intelligence as he had been under Walsingham. His position, however, was different: now he was more in charge, and occupied with the administrative problems of directing a network of political 'intelligencers'. Moreover, Bacon had become a principal confidant and advisor of Essex. Even if not employed by the Crown itself, he at last was engaged in giving counsel, and if Essex's political star could rise by his help, then the chances were strong that Bacon, too, could achieve a senior place in the state.

Not only the personal ambitions of Essex and Francis Bacon and the high politicking of the royal court and council are of relevance to us; broader issues existed which by the early 1590s concerned the political classes generally and with greater immediacy.

The first issue that comes to mind is the war with Spain. The maintenance of garrisons in the Low Countries and fleets in the English Channel was hugely expensive for the English Crown. Yet it would be wrong to overestimate the anxieties and the economic dislocations provoked in England by the war itself. A considerable part of the political nation (certain Privy Councillors included) rejoiced in the joining and continuance of open battle against the popish enemy, and the destruction of the 'Invincible Armada' in 1588 was almost universally believed to be a miracle signifying God's blessing and protection of Protestant England. Although the Exchequer's outlays increased greatly, and although parliaments in 1587, 1589 and 1593 were asked to provide ever larger subsidies and great City merchants were approached repeatedly for loans, it remains true that a long war against Europe's greatest and wealthiest empire was successfully being financed by a small state on Europe's periphery. Despite occasional complaints by individuals, parliaments and Convocations voted grants of unprecedented size in these years. Three decades of peace had strengthened England's domestic and international trading links, and the revolts in the Netherlands had furthered London's position by emasculating Antwerp as the principal trading and financial entrepot in northern Europe. London's merchants had much capital to invest and to lend, and the outbreak of the war against Spain, being a tremendous stimulus for trade, provided them with even more. English gentlemen and City merchants could well afford a war, especially one run by so parsimonious a prince as Elizabeth.[14]

Of greater immediate concern than the costs of the war to the governing classes in these years was the 'dearth'. Elizabethan gentlemen were struck forcefully by the grim prospect that 'dearth' could easily provoke grave threats to the stability of the established social order.

After thirty years of successive good harvests, from 1585 onwards arable crops in general – not just the grain harvest – were damaged repeatedly and sometimes ruined by poor weather. 1585 was bad and 1586 even worse, and this was reflected in prices for grains, peas and beans vastly increased beyond those of the 1570s and early 1580s. The harvests in 1589 and 1590 were no better. Although the price of wheat rose only slightly, peasants and urban labourers ate little wheaten bread, and prices for barley, oats, peas, beans and other arable crops equalled or exceeded the previous highs of just two years before. Hay and straw were similarly dear and precious livestock became more expensive to maintain. The crop failures in the late 1580s were severe and they came as grievous shocks.[15] Farming had always been a precarious livelihood, but these were the first failures since Queen Elizabeth had ascended the throne. Neither gentlemen nor peasants seriously regarded them as the effects of poor weather; they were explained as signs of divine anger at human wickedness.[16] But whose wickedness? That of the prince, the bishops, the landlords, the individual, or of mankind in general? Not surprisingly, Justices of the Peace and clergymen redoubled their efforts to lecture the people about personal sin (and the dire consequences of stealing), and the common people tended to focus on the evils of supposed grain hoarders, 'forestallers' and 'enclosurers' rather than criticise their masters. Nevertheless, the governing classes' perennial fears of peasant unrest and insurrection became even stronger in the late 1580s. Not only were Jesuits and recusants perceived as trying to seduce the people from their allegiances, and radical Puritan 'hedge-priests' suspected of trying to subvert the natural hierarchical social order, but the Justices now were confronted by widespread crop failures, price inflation, hunger, unemployment, disease and misery. Local rate-payers were expected by Tudor governments to provide relief for 'their' distressed poor, yet suddenly they were faced with calls for material assistance on an unprecedented scale. The magistrates of Northampton in 1586 wrote to Privy Councillors exclaiming that 'our poore people are so hardilie distressed that we stand in great dowte of some mutinie or unlawfull attempte to aryse amongeste them, unles somme politique meanes be devised'.[17] The situation was perceived as growing very dangerous indeed. However wealthy the rate-payers in general, and however carefully managed were the monies spent on the war, local demands for paternal assistance might quickly exceed the willingness – and even the ability – of individual gentlemen to pay: the consequences of this could well be disastrous for the regime. It was the very stuff from which magistrates' nightmares were made, and they were desperate to find solutions.

The Privy Council was quick to respond with its characteristic weaponry: regulations by means of royal proclamation. A royal proclamation was a powerful instrument of conciliar government, which was especially useful during sudden crises. It was typically addressed to sheriffs and JPs, demanding brisker enforcement of particular statutes or announcing new laws and new punishments.[18] For example, in a proclamation of October 1590, 'the Queen's Majesty, understanding the likelihood of great dearth of all manner of grain if good foresight not be had', forbade any form of removal abroad of 'any wheat, rye, barley, malt, beans, oats, or peas'.[19] A series of proclamations were issued against 'vagrant soldiers' (i.e., demobilised troops, who milled around the towns rather than returning to their villages) and 'vagabonds' (largely, unemployed men on the move): either sort of 'masterless men' made local magistrates nervous.[20] The councillors also issued proclamations regulating prices and wages, and sent out in 1586 for the first time the so-called 'Books of Orders', which listed for the JPs just what they should do in response to plagues and dearth. 'Each codified earlier practices and imposed a new uniformity on local authorities.'[21] The proclamations (and their severe penal clauses) were reasonably effective in generating more activity among local governors, yet they were not assumed to be cures for dearth *per se*, but to be directives on how to administer for its symptoms.

Anxieties about the price inflation and its possible consequences added further weight to the argument of Lord Burghley and his 'peace party' among the Privy Councillors: a halt to the Spanish war would free not only money and men but food. Much grain was being diverted to feed sailors and soldiers, and its removal from the local markets aggravated the scarcity and prices, breeding resentments against the government and its officers.

Another solution to the problem of dearth was offered to the public in 1589 by a minor member of the royal household. Anthony Marten argued that the dearth proved the need for his fellow countrymen 'to be in continuall readinesse' for imminent judgement:

> Do not the trees, the plants, the hearbes and fruits, and all other sencelesse things of the earth, which should be for the comfort of man, most sensibilie shewe unto us by the decay of their strength, of their taste and vertue (which not long since they had in greater measure) that within a short time, that there shall be no more use of

> them? Nay, doth not the earth itself complayne, that she hath powred out of her bowells in a manner all the store of her treasures and commodities, wherewith she was fraighted for the use of man?[22]

Treatises like Marten's are evidence both of the seriousness with which the dearth after 1585 was taken by the 'better sort', and of their increased efforts to focus attention upon personal sin and the moral reformation of the individual. With Judgement Day nearly upon them, people clearly should be busy preparing for His Second Coming by repenting and cleansing themselves. As Marten put it: 'Wee must either bring foorth fruite, or els bee cut down and cast into the fire.'[23] Because they distracted energies from the business of salvation, complaints about the dearth, no less than acts of disorderly conduct or open rebelliousness, were sinful.[24]

Another potential response to agricultural failures was the promotion of industry and manufacturing – the 'commonweal' policies as advocated by Sir Thomas Smith and put into practice by William Cecil. Through no fault of its own, this approach was no longer a viable solution by 1590. This grand policy had produced considerable successes (as well as some notorious failures) but it was a slow and long-term programme for creating domestic manufactures, employment and wealth, and before the dearth had begun it was being sacrificed to the short-term interests of money-hungry courtiers. One of the consequences of the growth of central government under Elizabeth was the growing number of officials and courtiers with claims upon the royal purse. Expenditures often exceeded current income, and in order to finance the increasing expense of governance, the Crown was forced on more than one occasion to sell estates and concessions which were part of the royal patrimony. Just like their medieval predecessors, Elizabethan parliaments regarded the costs of governing the state as a royal responsibility rather than a general one, and they expected the Crown 'to live of its own'. The state did not pay salaries of consequence to anyone, expecting office-holders to charge fees instead and to accept 'gifts'. By the mid–1580s, the system of protective patents for fledgling industries had fallen victim to the Crown's need to reward courtiers and officials, many of whom were rewarded with patents of monopoly over certain commodities or manufactures. This provoked great resentment among aspiring inventors and manufacturers ('projectors') as well as among established merchants and traders, for 'monopolists' would prosecute against a new or variant industrial process (which the old patent system had been designed to encourage), eliminate existing competition within an industry, skim the profits for themselves and 'restrain the liberties [to trade] of the subject'. Monopolists and patentees could use their royal patents to run roughshod over the wishes and jealously guarded authority of county magistrates, often with the connivance of Privy Councillors.[25] Conciliar support for new industrial and manufacturing 'projects' dwindled as the uproar against monopolies grew. Burghley continued to gather letters and petitions about new projects, but even he, who had promoted them vigorously for so many years, now accepted that the political climate was extremely unfavourable, and he stopped supporting them. Resentment appears to have been particularly strong among 'Godly' gentlemen and merchants.[26]

From about the mid–1580s, there appeared an increasing number of printed tracts and books which spoke the language of the earlier 'commonweal' programme and which presented their readers with information with which to 'improve' themselves. This was not to do with moral improvement *per se* (in the way Marten's tract had been) but they were concerned to promote material self-government or independence. The audience to whom these works were addressed was the landowning gentlemen and yeomen, and prominent among both authors and audience were the 'Godly'.[27]

Most of these works deal with mundane domestic issues, such as: remedies for cattle;[28] machinery for fishing and for trapping vermin;[29] medicines for various complaints;[30] how to build and stock fish ponds;[31] kitchen recipes;[32] the advantages of certain forms of coal, or fertilizers;[33] and the like. Some contain descriptions or allusions to new machines or 'devices'.[34] All these authors protested that their purpose was 'the benefit of the commonweal' (but who would say otherwise?) and many of their remarks suggest very strongly that the activity of giving away the fruits of their experience, of their 'experiments' (as they called them) or of their reading in foreign authors was pursued as a religious duty. As one man wrote: 'I ende nothing doubting but that all those which receive commoditie and ease by others travailes and labours, will not be unthankfull for the same but use them to God's glorie.'[35] In

1587, Leonard Mascall argued that he 'was ever desirous to further his countrie and common wealth, and said, that daie was evill spent, that he did not therein do good to his common wealth'.[36] John Taverner, who had 'a desire by all honest meanes possible, to benefit this my native countrie of England', explained his motives as 'seeking to shun that Monster Idleness'.[37] Hugh Plat protested:

> I will undertake to relieve poor Lazarus, who neither as yet, with the saltish drops of his body, is able by his labour, nor yet by the bleeding tears of his soul is able by his misery to procure sufficient maintenance the one way, or compassion the other way, to relieve himself, and his distressed family.[38]

Yet this 'undertaking' was not primarily in aid of charity for the poor and hungry (although assisting them successfully would be a clear sign of one's 'divine election'). Hugh Plat explained bluntly to his readers why he wanted their investment in a 'project' of his; he listed the reasons in order of their priority:

> First, their own safety which in these late years of dearth hath been a little touched, and more feared, and we have an old proverb that hunger breaketh through stone walls. Secondly, a sufficient employment of all such poor as are willing to work, and all excuse of idleness removed from lazy and loitering rogues. Thirdly ... the gain that shall hereby grow unto the rich ... Fourthly, the avoiding of many petty charges. Fiftly, the manifestation of a trew and charitable faith in relieving the extreme wants and miseries of their poor, and needy neighbors. Sixtly, the great encouragement of the Author ...[39]

Growing in the later 1580s was a stubborn opposition (which included many of the 'Godly') to an increasing number of Crown-imposed monopolies, with their far-reaching controls over local economies and local authorities. The London printing presses were producing large numbers of writings about domestic and estate husbandry, written by and for Puritan householders and gentlemen, for whom engaging in the activities suggested in these tracts was a religious duty. These two phenomena are not unrelated; both are manifestations of a 'Godly' sub-culture which was taking firm root within later Elizabethan society. It deserves our attention, since in writings of 1592, in which he first discussed natural philosophy, Francis Bacon was reacting sharply against this sub-culture and its way of knowledge.

With the evident failure – given the open hostility of the queen, her archbishop and many Privy Councillors – of the political movement for a greatly reformed Church of England, most of the 'Godly' chose to pursue 'reformation' at a local community level. 'It was time to turn with vigour to the reformation of towns, parishes, families and individuals, to be lost in the warfare of the spirit. Hence the paradox that the miscarriage of the further reformation [of the Church] coincided with the birth of the great age of puritan religious experience.'[40] The cradle for this Puritan religious experience was the select, 'voluntary' community of the 'Godly'. Many Puritans separated themselves from their 'lukewarm' neighbours and sought out the company and the approval of like-minded believers in other parishes as well as in their own:

> The puritans who composed the godly community were mentally and emotionally separated by their radical estrangement from conventional society and its mores and recreations, and by the fervour and strength of their own exclusive fellowship.[41]

One of the clearest manifestations of this 'separateness' and of the growth of 'voluntary communities' was the increasing attendance upon preachers and lecturers outside the established parochial system. The lecturer typically was either an under-beneficed and 'deserving' parish minister, one of a 'combination' of divines chosen to give a rotation of sermons, or a 'running' (i.e., itinerant) preacher, who travelled between various Puritan gatherings. Most of these men were the occasional employees either of town corporations or of the 'Godly' members of particular congregations.[42] Such lecturing was unauthorised by the Church hierarchy – even if certain bishops turned a blind eye to it – and the religious instruction delivered in them often ran counter to that handed down by 'approved' authorities.

These communities were not all-embracing ones: they were the exclusive preserve of the economically independent, and 'puritans knew that "the multitude", "the great unjust rude rabble" was not of their side'.[43] When Puritans spoke of the 'people' or the 'commonweal', they were less likely to be referring to all Englishmen or the national 'weal' than to the 'enlightened, responsible, dynamic, the spiritually industrious sort of people': that is, themselves alone and a 'commonweal' of the 'Godly'.[44] For example, Leonard Mascall did not speak of the 'commonweal of *all* England'; he referred instead to *his* 'countrie' (i.e., county) and to the furtherance of *his* commonwealth.

The 'voluntary' nature of the select communities was mirrored not only in the refusal of the 'Godly' to live entirely within the confines and expectations of the local parish or town society, but also in their 'voluntary' attitude to knowledge. We can see these attitudes to knowledge reflected in the 'gadding' to lectures: a voluntary harkening to the knowledge of the lecturer, who contributed his experiences to his auditors. If the lecturer was deemed insufficiently 'edifying', he could be replaced or superceded. Books written by and for the 'Godly' display the same convictions. Principal among these was the huge outpouring of religious books, particularly the thousands of penny or two-penny catechisms. 'These were not so much books to be read *as tools to be applied* by householders who took their religious responsibilities towards their children and servants seriously.'[45] 'Paracelsian' medical and alchemical writings in English flourished from the mid–1580s, full of simple medical and chemical recipes and advice, and their popularity among Puritan Englishmen was due largely to the uncompromising Protestant reforming ideology which permeated the 'chemical philosophy' of Paracelsus and his English champions. A considerable part of the 'chemical philosophy' was devoted to denigrating traditional intellectual authorities (e.g., the 'Ancients' and the 'Schoolmen') and to singing the praises of direct 'experience'. It carried forward, in other words, much of the religious reformers' assault upon the ingrained assumption that 'true' knowledge required the mediation of experts.[46] The 'how-to-do-it' tracts mentioned earlier are explicitly 'instrumentalist' in their attitude toward knowledge. Their authors make it clear they desired to volunteer their experiences about the exploitation of nature in order that the 'Godly' reader could be 'edified' by them.

A variety of evidence suggests, then, that the 'Godly' believed 'true' knowledge must be 'edifying': it must be constructive and useful in the unceasing battle to save the soul of the earnest believer. Such an attitude towards knowledge was an instrumentalist one; with salvation so desperately uncertain, any knowledge which did not appear to have a direct bearing upon the central struggle was of much less worth and of much less legitimacy than that which could assist the anxious Puritan. Furthermore, for the 'Godly' true knowledge was something found publicly and through personal participation and contribution rather than something privately determined and cautiously distributed by an especially privileged elite. Finally, the 'Godly' were convinced they were able individually to recognise 'true' knowledge: it required, they believed, no learned mediation and explication.

The Puritan retreat into select communities was viewed with deep suspicion by many members of the established Church and state. The 'Godly' were fully aware of the alarm their actions provoked: 'Is it not a factious thing', inquired the devil's advocate in a Puritan dialogue of 1584, 'that the people are affectioned to some men, and care not for hearing others?'[47] Bishop Cooper, in his anti-Marprelate treatise *An Admonition to the People of England*, argued bluntly that the Puritan lecturers were dangerous to the *political* order:

> When there were fewer preachers and less teaching by great odds than of late years hath been, the people did not revolt as now they do . . . in divers places where there hath been often preaching, and that by learned and grave men, there have been many that have revolted, and little good affection declared among the residue.[48]

As could be expected, Dr Bancroft and Archbishop Whitgift were ever willing to denounce 'the enemy within' and the evils of religious 'faction'.[49] In his 'Advertisement Touching the Controversies of the Church of England', Francis Bacon declared, 'some will no longer be *e numero*, of the number', and he condemned as false the knowledge of the 'Godly':

> [They] think themselves led by the wisdom which is from above, yet I say with St. James, *Non est ista sapientia de sursum descendens, sed terrena, animalis, diabolica: ubi enim zelus et contentio, ibi inconstantia et omne opus pravum.*[50]

The Puritan voluntary community flew in the face of the official Tudor ideal of an all-embracing, national community – the common weal of all English people – bound together within a single state and Church and within an all-embracing hierarchical structure. In the reformed England of Cromwell, Queen Elizabeth, Sir Nicholas Bacon, Burghley and Whitgift, there was absolutely no place for independent, aloof communities composed solely of 'equals'. Such a 'factious' desire by members of the 'better sort' was seen as irresponsible in the extreme. The ruling elite of Elizabethan England fully believed Thomas Cromwell's argument that 'England is an empire ... governed by one supreme head and king ... unto whom a body politic, *compact* of all sorts and degrees of people ... be bounden.'[51]

By 1591, Essex's flirtation with Puritan activists was over, and he would not have required his newest counsellor's advice to see the political wisdom of distancing himself from them. Clearly no one could protect them from the repression mounted by Whitgift and Bancroft in the early 1590s. From September 1590 onward, High Commission and Star Chamber trials and the lengthy imprisonments of leading Puritan controversialists smashed the radical movement beyond repair. The 'old days' and the old political alignments were being quickly swept away.[52] By this time as well, Francis Bacon was organising on Essex's behalf an intelligence network. That the command of such a network was deemed a prime political asset is illustrated by the fact that young Robert Cecil, as ambitious as Essex of a Privy Councillor's place, was now building an intelligence service of his own. The first skirmishes in a personal rivalry which soon after became a bitter feud engulfing the royal court were fought in the competition between Essex and Cecil for control of reliable 'intelligencers' and their information.[53]

It would be wrong to assume that Francis Bacon was completely occupied as an advisor to Essex: after all, Essex was not always in London where Bacon spent nearly all his time, and Bacon's attendance both at court and to Essex's affairs did not break his habitual routine of studying at Gray's Inn. Yet by 1592, when the Essex–Cecil rivalry began to divide the court, Bacon appeared sufficiently attached to Essex that Lord Burghley wished to be informed of his nephew's political allegiances and intentions. Bacon responded to his uncle's queries in a letter which is highly revealing of how he saw his political situation, his ambitions and his public duties.

Bacon's letter declares his determination to maintain a political independence despite Burghley's pressures to commit himself openly to the Cecils. His opening remarks in particular are full of cool reminders of the distance that separated them and of why

Bacon was likely to maintain it (namely, Burghley's failure to help his nephew's career):

> With as much confidence as mine own honest and faithful devotion unto your service [!] and your honourable correspondence unto me and my poor estate [!] can breed in a man, do I commend myself unto your Lordship.[54]

Bacon went on to characterise Burghley as 'the Atlas of this commonwealth, the honour of my house, and the second founder of my poor estate' – this latter conceit hinting that his powerful uncle, just like his father, had failed to provide Francis with the economic independence he felt himself due. This complaint notwithstanding, Bacon promised to resist any temptations to interfere with Burghley's political designs or to compete for place against men he might wish to advance:

> Perhaps you shall not find more strength and less encounter in any other. And if your Lordship shall find now, or at any time, that I do seek or affect any place whereunto any that is nearer unto your Lordship shall be concurrent, say then that I am a most dishonest man.[55]

Bacon was at pains to emphasise his loyalty and his desire to serve the Crown: 'I ever bare a mind (in some middle place that I could discharge) to serve her Majesty', since he was 'a man born under an excellent Sovereign, that deserveth the dedication of all men's abilities'. However, he made it very clear to Burghley that both how he wished to serve and just what he wished to do in the queen's service were unorthodox. *How* he wished to serve was 'not as a man born under Sol, that loveth honour; nor under Jupiter, that loveth business (for the contemplative planet carrieth me away wholly)'.[56] *What* Bacon wished to do as a counsellor for the queen was to administer a wholly novel office of state – 'all knowledge':

> I have taken all knowledge to be my province.

This was a declaration of great significance. Bacon had assured Burghley that he was uninterested in the usual scramble for posts (since he loved neither 'honour' nor 'business' but 'contemplation'), yet he also declared that he intended to pursue royal service 'in some middle place'. This 'middle place' was the *provincia* of 'all knowledge'. *Provincia*, or 'province' as Elizabethans rendered the Latin, referred both to an administrative unit (not a geographical one) and to a particular sphere of duty assigned to the official who governed it.[57] In short, Bacon had decided that 'knowledge' should be a *provincia* of the state and that he should be its governor. As a result of these convictions, Bacon assigned himself a brief:

> If I could purge it [the 'province'] of two sorts of rovers, whereof the one with frivolous disputations, confutations, and verbosities, the other with blind experiments and auricular traditions and impostures, hath committed so many spoils, I hope I should bring in industrious observations, grounded conclusions, and profitable inventions and discoveries; the best state of that province.

This may have sounded as odd to the practical Lord Burghley as it does to modern ears, but it followed not unreasonably from the conclusions about knowledge which the aspiring statesman had reached by the time of his 'Advertisement Touching the Controversies' of 1589. Bacon believed the first task was to conduct a 'purge' of 'rovers' who took 'spoils' in his 'province'. These 'rovers' were of two sorts, each of which wielded very different weaponry. The first ('with frivolous disputations, confutations, and verbosities') may be taken to be the university 'Schoolmen' with their dialectical methods. The second sort ('with blind experiments, auricular traditions, and impostures') are 'Hermetic' magicians, and the Paracelsian 'chemists' and 'Godly' authors whose works we met above. Certainly, Bacon's disparagement of traditional philosophic discourse echoed a complaint as common by the 1590s among counter-reformation Catholic scholars as well as among reforming Protestant ones. His criticism of the 'chemists' and the 'Godly' writers, on the other hand, stemmed from his suspicions of 'sects' in general and 'voluntary' unmediated knowledge. Bacon clearly was alarmed by the Puritan insistence upon unmediated distribution of information:

> Another point of great inconvenience and peril, is to entitle the people to hear controversies and all points of doctrine. They say no part of the counsel of God is to be supressed, nor the people defrauded: so as the difference which the Apostle maketh between *milk and strong meat* is confounded: and his precept *that the weak be not admitted unto questions and controversies* taketh no place.

He similarly attacked Puritan select communities

> although they have not cut themselves off from the body and communion of the church, yet do they affect certain cognizances and differences, wherin they seek to correspond amongst themselves, and to be separated from others. And it is truly said ... *there be as well schismatical fashions as opinions.*[58]

It is true that Bacon's desire to 'bring in industrious observations, grounded conclusions, and profitable inventions and discoveries' sounds very similar to those voiced by Puritan authors of 'how-to-do-it' books. Nevertheless, the intent is utterly different: Bacon is speaking of a state-led and state-controlled knowledge rather than a private one. By 'profitable', he meant 'profitable to the whole state' and not 'profitable' (i.e., 'edifying') to 'Godly' individuals alone. Bacon now thought the means for arriving at the 'best state of that province' were 'industrious observations' (supposedly foreign to the 'Schoolmen'), 'grounded conclusions' (unlike those of the 'Godly' who, in Whitgift's phrase, preferred 'opinions' to 'sure knowledge and certain proofs') and, following from these two activities, the achievement of 'profitable inventions and discoveries'. The attainment of this 'best state' Bacon believed to be '*philanthropia*' and, true to the 'commonweal' ideals and policies of his father, Burghley and others, Bacon confessed that his philanthropic *provincia* was 'so fixed in my mind as it cannot be removed'.

Bacon's letter provides little suggestion that he had many ideas of exactly how to proceed in this worthy task, but he was certain about general principles. He was captivated by a belief that knowledge could be harnessed to the advantage of the state as a whole *if it was centrally governed and organised*:

> I do easily see, that place of any reasonable countenance doth bring commandment of more wits than a man's own; which is the thing I greatly affect.

A 'place of reasonable countenance' meant a status or position which was respected by other men, and Bacon wanted more than this alone: he specifically desired a place with the 'commandment of more wits than [his] own'. Bacon had decided that the *philanthropia*, the achievement of the work, would require not only the intellectual labours of many men, but that he must have direct control over them. In other words, Bacon envisaged a hierarchical division of intellectual labour analogous to that which currently existed in the departments of state. His *provincia* 'all knowledge' – was to be organised just like any other principal instrument of royal governance and the royal will.

By the autumn of 1592, Essex's political star was rising rapidly at court. The annual celebrations of the queen's accession (17 November) provided a splendid opportunity to emphasise his ascendant position as the courtier-favourite of the queen. Accordingly, Francis Bacon prepared an entertainment, entitled 'Of Tribute, Or Giving what is Due'.[59] In this 'device' a princely figure asked his three companions each to 'do honor to that which he estemeth most and can most worthilie praise'. The first praised 'the worthiest vertue' ('fortitude'), the next man 'the worthiest affection' ('love') and the third praised 'the worthiest power' ('knowledge'). The companions then demanded a similar labour from their leader (probably Essex) who in a lengthy oration praised 'the worthiest person' ('his Sovereign', Queen Elizabeth). The entertainment was elegant, learned and devoted largely to the final speech in celebration of the queen, a paragon among rulers. 'Of Tribute' was an elaborate dramatic conceit with which Essex and his friends could entertain the court and flatter the queen on her special day. Yet it was more as well.

'Of Tribute' relied heavily upon a sixteenth-century humanist tradition of courtly literature known as 'mirrors for princes'. In such works, the author discussed the moral virtues which the good prince should have, including his youthful attainment of them by (humanist) education, his exercise of them as ruler, and the beneficial effects they (and he) would have upon the kingdom.[60] 'Mirrors for princes' were very well known by the 1590s; although he had put this didactic prose genre into a quasi-dramatic form, Bacon's audience would have recognised 'Of Tribute' as standing within the genre. The first two speeches in 'Of Tribute' were discussions of 'fortitude' and 'love' found, respectively, in Sir Thomas Elyot's *The Boke named the Governour* (1531) and Castiglione's *Il cortegiano* (1514), both of which were very popular in Elizabethan England.[61] There were further reasons why Bacon should have turned to these texts in particular. *The Governour* was the only 'mirror' to be written by an Englishman, and Elyot's book had been widely adopted as a sound guide by subsequent Tudor magistrates and 'governours'.[62] Castiglione's *Il cortegiano* had been translated into English in 1561 by Sir Thomas Hoby, Elizabeth's ambassador to France, and we can expect that Francis Bacon was familiar with *The Courtyer*, since Sir Thomas was one of his uncles.

His general familiarity with 'mirrors for princes' notwithstanding, Bacon had introduced something unlooked-for into 'Of Tribute': namely, the speech 'In Praise of Knowledge'. 'Mirrors' were devoted to the right conduct of the prince and his counsellors, that is, to *moral* philosophy. Bacon's 'In Praise of Knowledge' is unexpected in a 'mirror' *not* because 'knowledge' was unmentioned in these works (quite the contrary: an understanding of ethical behaviour was considered profound knowledge), but because Bacon praised *natural* philosophy. It was completely typical of sixteenth-century moral philosophers to argue that 'fortitude' and 'love' were crucial attributes of the great and good prince, yet Bacon was arguing that natural philosophy (and a strange, fantastic natural philosophy at that) had an equal right to be revered and to be pursued by the good prince and his companions. This was more than a witty twist to a familiar genre. It was impassioned advocacy by a man with a novel belief 'so fixed in my mind as it cannot be removed', as he had confessed several months earlier.

In this speech, Bacon made bold assertions about the knowledgeable man and the nature and uses of 'knowledge'. The speaker at first seems to be a traditional 'melancholic' scholar, deprecating the 'noyse and tumult of vaine and popular opinions' and repeating time-honoured saws in favour of the detached life of the intellect:

> Is there then any such happines as for a mans minde to be raised above the confusion of things, where he may have a respect of the order of nature and the error of men?

True happiness, then, lies in being transported by the intellect above the mundane, and gaining thereby a 'God's-eye-view' of both nature and men: a stock argument indeed, at least as old as Plato. Yet now Bacon overthrew expectations by proposing an addition to this utterly conventional proposition. In doing so, he fundamentally altered the purpose of the 'contemplative life':

> Is there but a view onelie of delight and not of discoverie? of contentment, and not of benifitt? shall we not discerne as well the riches of natures warehouse as the beautie of her shopp? Is truth barren? shall we not therby be able to produce worthy effects and to endow the life of man with infinite commodities?

The 'delight', 'contentment' and spiritual enlightenment of the individual is not the sole purpose of intellectual inquiry. Bacon argued that the man of knowledge must also accept an active responsibility: 'to endow the life of man with infinite commodities'.[63]

So to redefine the role of the man of knowledge would have been regarded as extraordinary by Bacon's audience, for whom such a figure would be not only unfamiliar but may well have seemed vulgar (particularly when compared, for example, with the high-minded moralists of the contemporary French academies). Bacon's work of advocacy was going to be difficult (he had, so to speak, an unknown 'client') yet we can see why Bacon thought his arguments reasonable. Capable of envisioning himself as the statesman of the *provincia* of knowledge, Bacon could also envisage what his subordinates would be doing: namely, bringing in 'industrious observations, grounded conclusions, and profitable inventions and discoveries'. Moreover, he deliberately spoke here not about 'endowing men' but about 'endowing the life of man'. This phrase was not an euphemism for 'seeking to improve men's inner virtues'; it was a direct statement about material commodities. Finally, that the 'endowment' of the life of man with commodities should be a goal for learned men was a conclusion which fitted comfortably with Bacon's prior commitment to the old 'commonweal' goal of national material prosperity.

Bacon had more surprises for his audience: the 'knowledge that is now in use', he declared, was insufficient for this task of 'discoverie', 'benefitt' and 'commodities'. Expanding upon remarks he had made to Burghley about the two sorts of 'rovers', he declared:

> All the Philosophie of nature which is now receaved is eyther the Philosophie of the Gretians or that other of the Alchimists. That of

> the Gretians hath the foundations in words, in ostentation, in confutation, in sectes, in Auditories, in schooles, in disputacions ... That of the Alchimists hath the foundation in imposture, in auriculer tradicions, and obscuritie ... The one is gathered out of a few vulgar observacions, and the other out of a few experiments of the furnace.

Significantly, Bacon had moved from speaking of 'knowledge' *per se* to 'the Philosophie of nature' and by doing so, he was implicitly redefining the role and purpose of natural philosophy. Accordingly, Bacon felt able to suggest why 'great reputed Authors' and the men of the universities had been so wrong about 'the philosophy of nature': 'alas they learne nothing there but to beleeve'.[64] The result had been the marriage of 'the minde of man' to 'vain nocions and blind experiments'. He wished instead to see 'the happie match between the minde of man and the nature of things'. It is notable that Bacon's man of knowledge and (*a fortiori*) his 'philosopher of nature', despite his new and civil duties, remained a man who contemplates, whose mind is 'raised above the confusion of things' and thus has a 'viewe', and who 'discernes'. He certainly is not an 'empirical' investigator of any sort; he remains a cogitator, pondering until truths become clear to him.[65] Although Bacon offered no specific suggestions as to how this 'happie match' might be achieved, he had *already* formed an opinion of what 'the posteritie and issue of so honourable a match may be':

> Printing, a gross invention; Artillerie, a thing not farr out of the way; the needle, a thing partly knowne before: what a change have these 3 made in the world in these times, the one in the state of learning, the other in the state of warre, the thirde in the state of treasure, commodities and navigation. And these were as I say but stumbled upon and lighted on by chaunce.

By the 1590s, it was usual to praise printing, artillery and the magnetic needle as evidence of a new European intellectual virility, a modernity which could challenge the 'Ancients'. Bacon chose to deny this comfortable claim.[66] It was notorious that 'printing', 'artillery' and the 'needle' were 'stumbled upon' by artisans rather than scholars (that is, they were *not* the 'issue' of the learned), yet Bacon argued (rather against the run of evidence) that a 'happie match between the minde of [his philosopher of nature] and the nature of things' would produce similarly 'worthie effects'. This argument seems to have depended less upon observation than upon the wishful thinking of a man with a prior commitment to the importance of producing 'worthie effects' or 'commodities': namely, a 'commonweal' statesman. This commitment also carried a powerful conviction about the importance of organisation and direction of affairs. Any true 'modernity' or 'reforms', as Bacon understood them, would come through prudent and vigorous governance from the top.

Bacon concluded his speech by arguing that 'hid in knowledge' lay 'the soveraingtye of man':

> Therfor no doubt the soveraingtye of man lieth hid in knowledge, wherin many things are reserved which kings with their treasure cannot buy, nor with their force commaund: their spies and intelligencies can give no newes of them: their seamen and discoverers cannot saile where they grow. Now we governe nature in opinions but are thrall to her in necessities. But if we would be led by her [in] invention we should command her in accion.

Bacon's language here, naturally, is explicitly political. He speaks of being a 'thrall', of 'governing', 'commanding' and of gaining power over the natural world. He already had argued that his privileged knowledgeable man, who had 'risen' to a 'respect' (i.e., a 'prospect') of both the 'order of nature' and the 'error of men', could produce 'worthie effects' (i.e., 'commodities'). Bacon now further equated this 'posteritie and issue' with 'the soveraingtie of man'. But what was meant by the enigmatic phrase 'the soveraingtie of man' (other than the fact that it decribes a political relationship) – especially as the context was a celebratory 'device' before the queen and court? It might simply be a romantic phrase suggesting the 'power of mankind over the natural world', but there may be a more concrete and local reference: namely, 'the English monarch's pre-eminence, supremacy or dominion over men'. Such a possible reading is supported in the lines that follow, which contrast the potency of 'knowledge' with what Bacon claims as the relative impotency of the means by which contemporary sovereigns were building states and empires. The implication here is that Bacon felt a principal use of this novel instrument of state was state-building itself.

Comparing 'Of Tribute' with other 'mirrors', we find that Bacon had added something he called 'knowledge' to the traditional 'fortitude' and 'love': all three, he claimed, should be pursued by the prince and his statesmen. Yet knowledge about the self has been displaced here by knowledge about the natural world: replaced, in fact, by a natural knowledge which had the same 'active' values and public ends as did the commonweal programme of his father's generation. The values and ends of that mid-Tudor programme were now elevated to the status of the values and ends of nothing less than a *philosophy of nature* by none other than a scion of the mid-Tudor governing elite.

But what, in substance, was this new philosophy of nature? Another courtly 'device' provided Bacon with a new opportunity to advocate his philosophical office of state. The opportunity came during the protracted Gray's Inn revels of Christmas-time 1594–5, when on a particular 'Grand Night' many of the Privy Councillors, great nobles and royal courtiers were entertained by the lawyers.[67] Appropriately, Bacon emphasised the role of Privy Councillors and the problem of giving 'good counsel'. He did this by presenting a group of councillors each of whom argued for a different grand policy for their prince and his state.

Bacon's device was well tailored to keep the interest of the audience of great councillors, court politicians and tyro magistrates. A prince figure asked six of his Privy Council to give their advice 'not of any particular action of our state, but in general of the scope and end whereunto you think it most for our honour and the happiness of our state that our government should be rightly bent and directed'.[68] In short speeches, the councillors argued for the 'Exercise of War', the 'Study of Philosophy', 'Eternizement and Fame by Buildings and Foundations', 'Absoluteness of state and Treasure', 'Virtue and a gracious government' and 'Pastimes and Sports'. The councillor advising the 'Study of Philosophy' implored the prince to exercise

> the best and purest part of the mind, and the most innocent and meriting conquest, being the conquest of the works of nature; making this proposition, that you bend the excellency of your spirits to the searching out, inventing, and discovering of all whatsoever is hid and secret in the world.

'The conquest of the works of nature' was now to be the goal, and

Bacon suggested the means by which this task should be accomplished:

> And to this purpose I will commend to your Highness four principal works and monuments of yourself: First, the collecting of a most perfect and general library, wherein whatsoever the wit of man hath heretofore committed to books of worth, be they ancient or modern, printed or manuscript, European or of the other parts, of one or other language, may be made contributory to your wisdom. Next, a spacious, wonderful garden, wherein whatsoever plant . . . may be with that care that appertaineth to the good prospering thereof set and cherished . . . rooms to stable in all rare beasts and to cage in all rare birds; with two lakes adjoining . . . so you may have in small compass a model of universal nature made private. The third, a goodly great cabinet, wherein whatsoever the hand of man by exquisite art or engine hath made rare in stuff, form, or motion; whatsoever singularity, chance and the shuffle of things hath produced; whatsoever Nature hath wrought in things that want life and may be kept; shall be sorted and included. The fourth such a still-house, so furnished with mills, instruments, furnaces, and vessels, as may be a palace fit for a philosopher's stone.

Four things must be noted here. First of all, these are indeed 'works and monuments': they require the participation and organisation of a great number of people and, presumably, for a great length of time. They are, in other words, *institutions*, and Bacon is advocating their establishment because he firmly believed institutional 'tools' are required for discovering knowledge of nature. Secondly, these institutions are devoted to collecting and then gleaning information, implying that the collection of writings and rarities is crucial for the enterprise. Thirdly, this is a royal enterprise: not merely because huge wealth is required but because Bacon desired these 'monuments' and the knowledge which they produce to be in the personal possession of the ruler of the state. Fourthly, a negative point: Bacon does not discuss how one proceeds from such man-made 'works and monuments' to the 'conquest of the works of nature'.

That Bacon should have a predilection for institutional solutions should not by now surprise us. At least from the time of his letter to Burghley, we may assume Bacon had been wrestling with the problems of exactly how his institutions would produce 'worthy effects' and 'infinite commodities'. By early 1595, he had convinced himself not only that collecting was necessary but it must be on a grand scale. Although such a spectacular effort does not appear to be a sufficient answer, in early 1595 Bacon had no further solution to the problem of 'conquering the works of nature'. Nonetheless, the instruments of his philosophical enterprise were to be royal institutions (that is, departments of the state) and however much the effect is *philanthropia*, the means for producing the knowledge which allows the 'conquest of the works of nature' and the 'soveraingtie of man' are owned by the sovereign. The product itself, the knowledge which results in 'infinite commodities', was to be generated on the monarch's behalf by a bureaucracy overseen by a loyal philosopher-statesman. We suggested that Bacon's philosophy of nature had the same 'active' values and public ends as the commonweal programme of his father's generation. The parallels are even closer: not only did Sir Nicholas and his youngest son share the same values and public ends, but they believed in the same *means* – the same Tudor bureaucracy.

Bacon's advocacy of a reformed natural philosophy was voiced during a period in which he was involved increasingly in matters of security and political stability and during which he articulated his fears about the social and political implications of the new 'voluntary' approach to knowledge of the 'Godly' gentry. His appeal was made, moreover, at a time when the venerable 'commonweal' programme visibly had faltered in practice. Bacon's advocacy in the early 1590s on behalf of a reformed natural philosophy was, in effect, a plea to resuscitate the 'commonweal' programme by reorienting and reorganising it.

4

Law

And to the text of the common law – *communis error facit jus* – one doctor says, *in favorabilibus*; another says, *facit jus*, subintellege, *dormire*: but the learned judges can awaken it when it pleases them ... we should consider the perils immanent in the present estate, who see in this time the desperate humours of divers men in devising treason and conspiracies – who being such men that, in the course of their ambition or other furious apprehensions, they make very small or no account of their proper lives. If to the common desire and sweetness of life the natural regard for their posterity be not adjoined, the bridle, I doubt, will be too weak: for when they see that whatever comes of themselves yet their posterity shall not be overthrown, they will be made more audacious to attempt such matters.

Argument in 'Chudleigh's Case' (1594), *Works*, VII, pp. 623, 633–4

How was Bacon's natural philosopher to achieve certain knowledge about natural things, and how was he to put it to use? The obvious (but unhelpful) answer is that he was to follow a novel procedure: but just what was this procedure, and was it novel? Like his aspirations for natural philosophy, Bacon's proposed procedures were utterly novel *for natural philosophers* but they were not novel in any absolute sense. Just as his aspirations for natural philosophy were adapted from the aspirations of his father's generation of Tudor statesmen, Bacon's procedures in natural philosophy were adapted by him from procedures in English law. Specifically, certain procedures in legal investigation and court trials, when linked with Bacon's own proposals for legal reform, not only exactly parallel his procedure for a reformed natural philosophy, but they were the model for it.

We have seen how fruitful it is for an understanding of his philosophy to remember that Francis Bacon was a Tudor statesman's son and to see what in particular this entailed for him. It is equally important to remember that he was an experienced and learned Elizabethan common lawyer: an appreciation of the workings of the law and the assumptions of the professional lawyers is essential for understanding Bacon. In the 1580s and 1590s, Bacon rose rapidly and high in the legal profession. True, it was in the middle years of King James' reign that Bacon won the great law offices (Solicitor-General, Attorney-General and then Lord Chancellor), but whatever the particular demands of these offices of state, they introduced Bacon to little about the actual workings of the law which he had not already experienced during the reign of Queen Elizabeth.

For Bacon, 'discovering' (i.e., uncovering) the truths of natural philosophy involved several stages, for each of which an exactly parallel feature existed within the law as he practised it or wanted it practised. The present chapter examines certain aspects of the Elizabethan law system with this theme of 'discovery' in mind; after all, only those legal processes and institutions which served Bacon as models for a reformed natural philosophy need concern us here.

The story of late medieval and Tudor English law is principally the story of the successes and developments of the king's justice. Despite opinion to the contrary (especially among seventeenth-century lawyers), England's law is 'common law' *not* because it had been identical with the ancient and local customs of the 'common people' but because it was general: it was the law dispensed throughout the land by the Norman, Angevin and Plantagenet kings. Wherever the royal court was to be found (and it moved ceaselessly), there the king and his closest advisors would hear and judge pleas, but during the later twelfth century, men to whom the king had delegated some of his authority to judge disputes began to reside at Westminster and they occasionally toured the countryside in his name as well. Here began the great royal law courts and the regular assizes in the countryside, and the permanent loss to local courts of much business and administrative authority.[1] The Angevin kings did not dispense a fundamentally different kind of law than did the local barons and dignitaries (although they always asserted the right to judge serious crimes – 'the pleas of the Crown'). They, too, were largely concerned with tenures, jurisdictions and obligations, but the king's authority and right to give justice necessarily surpassed that of any lesser lord (or so monarchs repeatedly claimed).

Disputants, for their part, were content to approach the king's courts for justice, rather than enter the court of a local lord and there perhaps, by means of a supernatural proof, be forced to await the verdict of the 'infallible judge'.[2] When disputants chose to appeal to a court, rather than engage in 'self-help', they were searching for an authoritative settlement, and in the twelfth century not only was the king's institutional capacity to govern steadily growing but his authority to do so was as well, relative both to local lords and to God. His great courts gave the king the machinery with which to rule and to judge, and his scholarly propagandists emphasised the semi-sacred status of the king and the divine legitimacy of his right to dispense justice and good government. The argument was that legitimacy and authority were conferred directly upon the king by God (for example, through the coronation unction), that a sacred duty had been placed upon him to right the wrongs of his people and that the royal common law did not supplant the laws of God but was fully in accord with them.[3] Weighed against this, the authority of local courts seemed inadequate and the capriciousness of 'battle' and 'ordeal' unnecessary; men took full advantage of the growing opportunities to appeal to the king's courts of law.

When England's kings created courts of law, they created also the new classes of royal government servants who worked in them: the judges, advocates and clerks. Professional royal justices, whose authority was derived solely from an office they held at the king's pleasure, consistently delivered the same justice in their master's court of law, and for their own guidance they referred to their own and other court-professionals' previous experiences and procedures there rather than defer to supernatural proofs. In other words, before the rational analysis of disputes became general in England, it was necessary to have royal judicial servants, permanently employed in discharging their king's responsibility for justice, and the institutional memory which followed from this permanence.

Men now could expect rational analysis of their disputes, and from courts possessed by the highest authority in the land. Men would inquire into the 'facts' of a given dispute, men would make decisions thereafter (based partly at least on their inquiries, and following the 'course' of the court), and not even the great barons could successfully ignore or overrule the commands of a royal court of law. These great changes occurred in the later twelfth and early thirteenth centuries, and 'not until that happened ... did we have a substantive system of Law'.[4]

We tend to think of 'courts' as institutions where adjudication about rights and obligations is conducted by lawyers, and where judges mete out punishments upon the guilty. It is difficult to recapture the sense of the term 'court' which was prevalent in late medieval and Tudor England, and to remember that 'courts' were not devoted solely to the adjudication of disputes. 'Court' was a general term, describing those formal and continuing arrangements by which gentlemen throughout the country conducted and oversaw business of almost any sort (both for themselves and others). 'To hold court' was to govern. The right to hold court certainly gave its owner the duty to dispense justice, if that was what the immediate situation required.[5] Feudal lords governed their lands and servants in seignorial courts. The ancient divisions of the country were administered by the borough, hundred and county (sheriff's) courts. The royal household had many different 'courts' beside the 'royal court' where the 'courtiers' roamed; even victualling the household was administered through the 'Court of the Green Cloth'. In short, the governance of the entire kingdom was conducted through 'courts'. The 'common law' exists as 'a by-product of an administrative triumph, the way in which the government of England came to be centralised and specialised during the centuries after the Conquest'.[6] The medieval royal courts of Exchequer, King's Bench and Common Pleas did more than 'give justice' to the king's subjects; they were the new, specialised institutions by which he ruled and administered his realm with an increasing authority. By mid-Tudor times, these courts had been supplemented by newer ones, such as Star Chamber, Requests, and Wards. These often have been called 'prerogative courts' in order to contrast them unfavourably with 'common law' courts, but this distinction is spurious and entirely misleading: 'the "common law" courts of the sixteenth century had been the "prerogative courts" of Angevin days'.[7] As a consequence of this administrative triumph, the nature and purpose of law and courts 'of law' were understood to be much broader than the business of settling disputes alone. Law and courts of law were for the governance of England.

When individuals first accepted the arbitration of courts for their disputes rather than engage in 'self-help', a very great step in communal and political organisation was being taken. Every civil society creates arrangements to do this, the aim (and the need) being to preserve peace and order in the community. The existence in Tudor society of a complex and formal machinery for pursuing suits and court trials, of elaborate legal doctrines, and of many hundreds of professionals (with highly specialised vocabularies and self-interests) whose business it was to act in or for courts, should not distract one from seeing that the maintenance of peace and order remained an essential aim of the law courts. Like all Tudor Lord Chancellors and Lord Keepers, Sir Nicholas Bacon emphasised exactly this theme when opening parliaments and 'charging' judges going on circuits. Ordinary barristers echoed their superiors: 'The common law was framed in favour of public tranquillity', declared William Fulbeck, 'and therefore the departure from it must needs be accounted odious.'[8] The point is illustrated also by considering what happened when a grievance was brought to a court trial.

Suppose that a neighbour objects to my pasturing sheep in a certain field; although I protest that I have a right to do this, he disagrees and builds a fence barring my sheep from the field. For both of us a very particular and unique dispute has begun, high emotions are involved, and it easily could provoke violence. But if I resort not to 'self-help' (e.g., intimidating him with a big stick) but to a lawyer, what happens? Our emotions still are running high, yet my lawyer chooses to construe our dispute as a question about (say) a local definition of pastureland – something I had barely considered. Since lawyers deal in points of law, and not in passions, every unique dispute brought to them necessarily will become converted into the categories and terms in which they deal. This squeezes out both the emotion and the particularity from our confrontation: 'our' dispute now has been rendered into the only form that a law court is able to resolve – a legal form, an artefact, a surrogate dispute conducted by lawyers about legal entities. The dispute has been taken out of our hands and scrutinised and pronounced upon by third parties whose behaviour and language we may little comprehend but whose decisions we accept (however grudgingly) as authoritative. My neighbour and I may continue hereafter to detest each other, but after the court's decision this particular occurrence of open and declared confrontation no longer remains.

It will be valuable to summarise (in a highly abstracted form) what a late medieval and Tudor court trial consisted of. Put in its simplest terms, in a court there is a judge, who acts on the king's behalf to see the law observed and justice done, and who gets aid from (and guides) a jury. In earlier centuries, a jury consisted of oath-takers who would swear to the innocence of one of the parties in a suit or to the truth of his testimony,[9] but by Tudor times the jury was a panel judging the credibility of testimony and delivering their verdict upon this to the judge. Rather than pursue their lawsuits personally, parties were represented by professional courtmen because the 'course' of the courts (and thus the manœuvering required to succeed in a suit) had become extremely complex. When a party complained and sought redress in a court alleging that another had infringed the law by a certain action, then a case had begun. Plaintiffs always alleged two things: that something had happened, and that it was an infringement of the law. Consequently, a suit required that two questions be answered to the judge's satisfaction: first of all, 'did it happen?' and secondly, 'was it an infringement?'

The Tudor legal term which referred to an 'event', 'deed', or a 'thing done' was the term: *fact*. Although today a 'fact' is taken immediately to mean something indisputable or certain, for the Tudor lawyers a deed would *never* be understood in this manner in the first instance. 'Facts' were continually being alleged by parties at law, and they had to be inquired after, further uncovered and analysed. The sense of *fact* most common in Francis Bacon's day, for example, is preserved in the legal phrase 'accessory after the fact', where 'fact' is an evil deed.[10] Elaborate procedural machinery was developed by the courts in order 'to establish facts'.[11] At the beginning of his suit, the plaintiff asserted that something had happened, but in order for his allegations of fact to become accepted as true, he needed to convince the jury and the judge that his allegations were 'facts' indeed.

The principal way 'to establish', or 'to prove' that one's allegations were true was to accumulate and to present the testimony of witnesses.[12] The witnesses would swear an oath to tell what it was that they knew, and the other party's advocate would try to undermine or supplant their testimony by questioning them or other witnesses.[13] The status and the repute of witnesses mattered a very great deal. For instance, if a man was present at an event, his testimony was of greater weight than if it was second-hand knowledge: he was logically a superior witness *qua* witness. In practice, the testimony of a landed gentleman would be of greater weight than that of a village labourer, a serving girl or a beggar: his word was accorded greater respect. The quality of the witnesses was actually of greater importance for the establishment of 'facts' than the quality of the parties themselves. The aim was to convince the jury (and thus the judge) by means of the quantity, quality and consistency of the testimony you provided that what you had alleged was true. To 'establish' a 'fact' (i.e., to settle to the court's satisfaction that a particular event had happened, and what it was), a party had to gather testimony, and then successfully 'prove' (i.e., 'test', 'argue for') it to the jury and the judge by means of rigid procedures for witnessing, argument and the assessment of testimony. Lawyers were required to follow accepted procedures and to build assent in order to 'establish facts' in a court. 'Facts' were legal entities; they were artefacts carefully constructed during the court trial of a lawsuit.

Disputes often were concerned solely with the question of whether a particular 'fact' had indeed occurred (e.g., given that Richard was stabbed: 'Was it *John* who stabbed Richard, as is alleged?'). If this allegation became 'proven', then it was usually obvious that an infringement of the law had taken place, and the final judgement would follow. In most suits, once the facts had become established, there was no debate about whether an infringement had occurred as well; the advocates and the judge could point to statutes or to previous judgements of the courts which would tell them what the law was.

Yet sometimes a dispute was not about the facts, but about whether or not an infringement had occurred. John had taken possession of lands once held by Richard's father, now deceased. One party would claim that the 'fact' was an infringement of a particular law, and the other would disagree, claiming that this law did not apply in the present case. Even further, there were occasions when appeals were not made directly to a known law, but rather to an interpretation placed upon known laws and older judgements, and this interpretation was claimed as support for one's insistence (or denial) that an infringement of the law had occurred. By the sixteenth century, the common law was regarded as encompassing more than the statutes, royal proclamations and the older judicial decisions; it was believed to be a huge body of law, including much which was unwritten. Moreover, it was believed to be structured in accord with reason, and because of this structure, when the known laws did not give clear indications of infringements, lawyers would appeal to what they took to be the implicit message of the common law. Since the common law was believed to be consistent, as well as rational in character, a lawyer was prepared to argue from other cases to the present one and, after examining those cases, to produce a statement about the law implicit in them in order to bolster his present claims. The lawyer's task in cases of this sort was to uncover what the law, although hidden, had always 'said' about this case. This will be explored more fully below, but we need first to consider more closely specific procedures by which testimony was gathered and facts 'established' in Elizabethan court cases.

By the later sixteenth century, the great royal Court of Chancery in particular had worked out elaborate procedures for gathering testimony in civil suits, and it was possible for a party at law to purchase the use of this Chancery process even when the suit was entered in another court (e.g., in King's Bench, Wards or in a local court).[14]

In Chancery, the establishment of facts usually was achieved by means of 'examinations'. After the plaintiff registered his grievance with the court ('exhibited his bill') and the defendant then exhibited his answer, a Chancery *subpoena ad testificandum* would be served on each witness named by either side, and men empowered by the court would begin the lengthy business of gathering their testimony. This could be done either 'in town' (e.g., in Chancery Lane) by the Chancery examiners themselves or 'in the country', for which a commission for examinations was awarded to the nominees of one or both parties.[15] Not surprisingly, Chancery examination, particularly examination in the country (which by definition was not supervised by regular officials of the court), became surrounded by a forest of rules designed to ensure the integrity of the procedure; yet the essential structure remained straightforward.

Each witness was asked to respond upon oath to a list of questions (a 'schedule of interrogatories') framed around the pleadings made either by the plaintiff or by the defendant (depending on whose witness this was). The relevant counsel had written out these interrogatories and signed the parchment, a Master in Chancery scrutinised the questions in the schedule 'for sufficiency', and then the examiners or commissioners began their work.[16] The number of questions varied greatly between one, a few and many dozens, but they always took a precise and formal shape. For example: 'Item, do you not know or have you not credibly heard or are you not fully persuaded in your conscience that it was the true intent, will, and meaning of the said Nicholas Bristowe, deceased, that ... ?' Such a question illustrates that a testimony was never expected to provide *certain* knowledge but probabilities only.[17] The witness responded orally but his answers were written down by the examiner's clerk (or one of the commissioners) in a manner as formal as the interrogatories themselves. The witness then read over his statement and requested alterations if he saw mistakes there. Such a 'deposition' was not identical with, yet faithfully represented, the words actually uttered by the witness in reply to the questions. All the depositions, together with the schedule of interrogatories, were bundled together and returned into the court, where they would be scrutinised again by a Master in Chancery.

The purpose for the strict formality of the questions and the recorded answers was fairness to both parties. Because each witness was asked the same questions in the same order and in the same form, and because each witness' answers were represented in the same form of words, proffered testimony which was extraneous to the substance of the original pleading could be identified and discounted, and the depositions then could be directly compared. If the Master was dissatisfied with a particular deposition, he could exclude certain remarks or order a *viva voce* examination to supplement an answer.[18] It was only after his careful sifting of all the depositions had been completed that their 'publication' was made. At last, both parties could see all the depositions and take copies of them in order to further the presentation of their suit in the court in which it had begun.[19]

Similar arrangements of interrogatories and depositions were employed in the Court of Star Chamber. The significant difference from the procedures of the Chancery was that in addition to the witnesses, here the defendants themselves were examined upon oath. In Chancery, the defendant's answer to the plaintiff's bill was given after swearing an oath and, because it met all the points in that bill, his examination was not thought necessary. In Star Chamber, where the Privy Council and Chief Justices sat to enforce the king's peace (plaintiffs routinely alleged 'riot and affray'), to enforce royal proclamations and to punish abuses of justice (e.g., perjury, influencing juries or contempts of court), it was thought necessary to examine defendants.[20] Although most Star Chamber suits were between private parties, it is not surprising, given the central interests of the court, that many suits were brought here by the Crown's law officers as well. For example, on 19 October 1597, in a suit against corn engrossers, 'Kooke, the Attorney-General, informed *ore tenus*, on the confession of the parties themselves ... these ingrossers were examined by Bacon, Counsel at Large to the Queen'.[21]

In the course of their ongoing concern with security and treasonable matters, the Privy Councillors authorised trusted commissioners to examine suspects.[22] Because these warrants were from councillors sitting as the Privy Council rather than as members of a formal court (with specialised officials, records, and fixed procedures), it is difficult to be certain how examinations were typically conducted. However, we can read in the Privy Council's register and in the state papers many letters ordering commissions and many letters from the examiners back to the council; these do allow us to say something about the examinations.

Suspects were brought to the Privy Councillors' attention by any number of means: by information provided by citizens, alert justices, common informers or through the networks of 'intelligencers' privately maintained by Elizabeth's councillors.[23] Once a JP or a Privy Councillor himself had heard the testimony of witnesses or informants, and was convinced some grave treachery was afoot, he would order the arrest of the suspects and would notify the council. All these suspects, then, would receive their examinations while in custody. The examinations followed the familiar structure of interrogatories and depositions, but unlike those examinations in ordinary lawsuits outlined above, the examiners would repeat the procedure again and again until completely satisfied they knew everything that had happened and the names of everyone involved. For example, suppose a man was arrested on strong suspicion of knowingly harbouring a seminary priest, who had been captured leaving his house. The information initially brought to the JP (and any papers and documents then found in the suspect's lodgings) would provide a basis for the first interrogatories put separately to both men by the examiners. The least of the examiners' concerns was to confirm that the prisoners were guilty of the crimes for which they had been arrested: the allegations upon which the JP had acted usually were sufficiently credible for that. They were anxious instead about a Catholic 'underground', and they wanted intelligence about any new plots and conspirators: what did these prisoners know of such matters? The depositions taken were analysed for inconsistencies, obvious evasions and contradictions, and then compared against each other. (If one of the prisoners had implicated someone else, then he, too, might be arrested, and drawn into the inquiry.) A new schedule of interrogatories was drawn up and the prisoners were examined again, and so on. Torture, in a judicial sense, was used sparingly, although suspects often were subjected to the terrors and calculated misery of precarious diet, isolation, vermin, damp and the likelihood of catching fevers.

Examiners given commissions to torture knew long patience might provoke confessions, but how could the integrity of these be known? For such elaborate procedures to be successful against unwilling persons whose mental and physical health frequently was deteriorating, the examiners needed to rely principally upon their analytic skills in framing probing questions and scrutinising the depositions. The words from a prisoner's lips were in themselves unworthy of the examiners' trust (even when the examiners were permitted to torture, and were then joined by skilled 'rack-masters' such as Thomas Norton, Richard Topcliffe or Richard Young).[24] As Francis Bacon remarked, 'in the highest cases of treasons, torture is used for discovery, and not for evidence'. This was widely understood by laymen as well:

> Although no man be tyed to accuse himself by any compact, yet in a publique tryall he may, by torture, be forc'd to make answer; but such answers are no testimony of the fact, but helps for the searching out of truth.[25]

What examiners did trust was the information finally resulting from their rigorous and repeated deployment of an examination procedure completely under their control: that is, they regarded as true the final account (i.e., the facts) of what happened, why, and just who was involved, which they themselves had generated from the many layers of testimony, questions, sifting, cross-referencing and analysis. When such a report of facts was presented at a trial of the defendants, irrespective of whether the defendants now chose to repudiate it, it would almost certainly be regarded by a court as compelling evidence. This may help to explain why these grisly assignments were usually given to men who had great respect for formal processes in general, the ability to reconstruct past deeds from often conflicting information, and experience in estimating the full extent of any criminal plot: namely, the law officers and a select circle of senior barristers trusted by the Privy Council.

A civil suit, once in the court room, opened with the plaintiff's declaration (or 'count'), in which he asserted the facts upon which he grounded his claim, and the suit then proceeded by a structure of allegations ('pleadings') made by the barristers acting for the defendant and plaintiff in turn until they reached 'issue'.[26] Sir Thomas Smith described suits in the following manner: 'part by writing, and part by declaration and altercation of the advocates the one with the other, it doth so proceede . . . till it doe come to the issue'.[27] The function of pleadings was to find or 'reach' an issue – any issue. It was the 'issue' which the jury were required to answer in favour of one party or the other, and to help them in their task they had any facts already established and the various allegations of fact in the pleadings and, perhaps, some guidance from the judge as to what was the law. The jurors 'determined' the 'issue' by finding the defendant either 'not guilty' or 'guilty', and the judge then passed sentence.

But what was an 'issue'? Like a fact, it too was an artefact of the court room. An 'issue' was reached whenever an allegation (or the assumption about the law which an allegation might contain) was directly challenged by the other party. For example, let us recall my dispute with my neighbour over the pasturing of my sheep. If in my 'count', I had declared that John previously had given me leave to pasture my sheep in that field, and John's answer was that he had not done so, then we (or, rather, our barristers) have reached 'issue', since a specific denial of an assertion had been made. The jurors would now be asked to determine this issue (i.e., 'did John give the permission, or not?'). Yet, perhaps John would not wish to 'join issue' on ground chosen by me; he might wish to introduce new information by which to exonerate himself, rather than straightforwardly deny my declaration. Therefore, John might reply, 'Yes, I had consented, but only because I was forced into it.' I then could choose to deny this directly (thus reaching an 'issue': was John's consent forced from him, or not?) or I could 'confess' to the new information (as John had done) but 'avoid' issue by introducing my own extenuating information. And so on. Lawsuits, then, involved the sparring of the barristers as to which possible issue they would finally 'join' upon, since each hoped to reach an issue advantageous to his client. The strategies governing the tactical deployment of the various sorts of pleadings were a central part of the barristers' trade, and they were known among the men of the Inns as the 'science of well pleading'. As Sir Thomas Littleton put it, 'know, my son, that it is one of the most honourable, profitable things in our law to have the science of well pleading in actions real and personal'.[28] Elizabethan common lawyers recognised, as had the lawyers of ancient Rome and the medieval Church, that their 'science of well pleading' rested squarely upon the ancient art of dialectic, a knowledge of what is likely or probable, or how legitimately to construct proofs by argument from probabilities rather than from demonstrable certainties (as the philosophers preferred). 'Dialectica est scientia probabiliter de quovis themate disserendi', remarked Sir Edward Coke, 'whereby it appeareth how necessary it is for our student.'[29] Yet whatever the issue reached in a lawsuit – no matter how seemingly marginal or far removed from the fullness of the original 'real' dispute – this alone was the question put by the judge to the jury for their determination.

Although it was a fallacy of reification and a professional myth, sixteenth-century common lawyers firmly believed that a huge substantive body of law, 'The Common Law', actually existed in its own right, encompassing any eventuality and holding answers for their every query. Of course, such a faith brought practical difficulties. Because 'The Common Law' was to be understood as an omnicompetent body of law, and since disputes without number were forever springing up – each one unique in its details – the great bulk of the law was necessarily unknown before particular court cases had been adjudged. On the one hand, the myth sustained the lawyers' confidence that professional inquirers into the law were indispensable for the common good, yet, on the other hand, their difficulty (like that of the priesthood of any oracle) remained that 'The Law' was not available to them in a single great 'book'. Nevertheless, by conscientiously 'studying the authorities' (i.e., reading texts) and by becoming 'learned in the law', lawyers expected to glimpse at some of the hidden pages.

In reviewing some of the methods and sources by which Elizabethan lawyers sought to discover their hidden common law, we face a problem of historical exposition. Legal historians can see that during the sixteenth century great changes were occurring in lawyers' assumptions about how to discover the law, yet it is doubtful many Tudor lawyers were aware of this. 'One of the difficulties in understanding common law change is precisely that it hardly ever represents anybody's mental process. It is the cumulative result of small interactions.'[30] At the risk of making these changes appear more clear-cut and more widely understood than they were, we shall sketch the situation which existed in the greater part of the century, and then discuss the new assumptions which were gaining acceptance by the end of Queen Elizabeth's reign.

Although in most Tudor lawsuits, the issue reached would be about facts, occasions arose when the issue reached was one of law. 'Issues of law' could be reached in several ways, the most typical manœuvre being the 'demurrer to the evidence': a party chose to accept the allegations of facts made by the opponent but denied the relevance of the point of law which the opponent (by pleading those particular allegations) implied was in his favour. A suit that had proceeded to a demurrer promised to become intellectually intriguing for the lawyers themselves. Any issue now reached would be framed with questions such as these in mind: 'What is the law on such-and-such a topic?' or 'What does the common law "say" about this point?' And not only during a trial of an 'issue of law' but also with new forms of delayed judgement following a trial (increasingly employed as the century progressed), Tudor lawyers and judges attempted to uncover answers to questions of this sort.[31] They did this through discussion and argument largely among themselves: during dinners with professional colleagues, at the 'moots' and 'readings' in their Inns, or (most formally) in the courts.[32] These issues, arguments or judgements often were of such interest to a novice, a serjeant or a judge that he took notes for his future reference. This was one of the ways a man became 'learned in the law', and the desire to become 'learned in the law' was the essential motive for the habit of making 'law reports', which provided lawyers with a means of remembering earlier discoveries of 'Law'.

English lawyers and judges always took notes as private *aides mémoires* on aspects of law cases that interested them. Yet most cases were not 'interesting'; that is, with regard to the technicalities of pleading or procedures, they were the same as a hundred others: a professional lawyer learned from them nothing new about his trade. But a novel problem over pleadings or an argument among the judges and serjeants about a point of law was likely to attract a note-taker's attention; nevertheless, if the case otherwise fell into a familiar pattern, no other proceedings (often not even the eventual outcome) would be recorded. After all, noting all the (predictable) moves in the progress of a typical suit would have been pointless labour for the lawyers.[33]

Lawyers often kept up their notebooks throughout their careers and referred to them in order to remind themselves and persuade others (if possible) about the arguments or procedures in earlier cases. The notebooks of retired judges and serjeants were much prized, and lawyers traded and inherited them, copying notes about old cases into their own books. With the advent of printing in the late fifteenth century, and when London printers saw that a market of eager customers existed for legal books, some of these manuscript collections quickly became available in printed form. The printers gathered together whatever old reports they could get, arranged the cases in them chronologically by law terms and regnal years (as many of the individual lawyers had done for themselves), and presented for sale a series of books containing a very large number of reports on cases for most law terms since the early fourteenth century. Lawyers still took notes and borrowed from each other, but purchasing these printed volumes (known, reasonably enough, as 'The Year Books') largely replaced hunting up one's own collection of older cases.[34]

The Year Books, and texts such as Sir Thomas Littleton's *Tenures* (*c.* 1481) and Sir Anthony Fitzherbert's *Graunde Abridgement* (1514), were regarded as reflecting the 'common erudition' of the generations of professionals at Westminster and the Inns.[35] 'Common erudition' was the term given by the lawyers to the group wisdom of their professional elite, and Tudor lawyers revered it; it represented what had been gleaned to date about the common law. When judges and serjeants met to discover the law,

> the law which emerged in the course of argument was not law laid down by the court so much as law which was accepted learning within the profession ... The judges were the chief repositories of this common learning, but what they said as a body was law mainly because it conformed to the reasoning of the little intellectual world of Westminster Hall.[36]

The lawyers believed the 'common erudition' was becoming a more refined and insightful body of knowledge about the common law. Sir Edward Coke declared that the young law student should first study the more recent reports: 'For the most part the latter judgements and resolutions are the surest and therefore it is best to season them in the beginning, both for the settling of his judgement, and for the retaining of them in memorie.'[37] Over the decades, the 'common erudition' was altered without these alterations ever seeming so sudden or so peculiar that they would not be viewed as reasonable refinements of older insights. We can see this optimistic assumption at work in the Tudor attitude to the Year Books and to ancient law treatises such as 'Glanvil' and 'Bracton'.

The Year Books were regarded as illustrations of past practices and understanding, rather than as sacred texts always to be followed. An old judicial decision that offended the present judges' sense of the 'common erudition' would be ignored by them as 'unlearned' and outside the common wisdom about the law.[38] Of course, a judge would not lightly ignore the learning gleaned from older reports, but there was no sense that earlier decisions found in the Year Books were so authoritative that present decisions must be bound by them. A remark of F.W. Maitland is as appropriate for 1500 as it was for the 'lawyer of Edward II's day', who

> regarded his copy of reported debates rather as we should regard a manuscript copy of precedents in pleading than as we regard our law reports. He studied it in order that he might be prepared to make the correct ... 'book moves' ... but he did not think that the Court would be bound to decide a given case in a given way, because he had a similar case in his book.[39]

Quite unlike the iron doctrine of later centuries, there was no conviction that 'precedents' must be strictly followed.

Similar attitudes governed the Tudor lawyers' opinion about the value of texts like Ranulf Glanvil's *De legibus et consuetudinibus Angliae* (*c.* 1187) and Henry de Bracton's similarly titled work of the 1250s. Despite their attempts to provide a summary view of the laws and customs of their time, neither text was much used in court-room argument in the late middle ages.[40] Their statements occasionally were introduced for rhetorical 'colour', and 'Glanvil' and 'Bracton' were therefore well known as 'ornaments of the law', but they were not seen to be authoritative in themselves. This opinion was confirmed for the Tudors by Fitzherbert, in whose *Abridgement* was a remark from a case in 1457: 'tout le court dis que Bracton ne fuit unques tenus pur auctor en nostre ley'.[41] In a case of 1587, 'it was well observed, that Glanvil, Bracton, Britton, and Fleta, may be vouched for antiquity, and ornament in cases where they concur with the later authority of law, and do not impugn the common experience and allowance in judicial proceedings at this day'.[42] The 'common experience' of the profession was taken as the most important source of authority. Tudor lawyers firmly believed it was by study and argument amongst themselves about the law that they would best sustain their ongoing miracle: the 'slow revelation and refinement of doctrines essentially immutable'.[43]

Not only did Tudor common lawyers believe 'The Common Law' encompassed every eventuality but they believed as well that the law was rational and that it had a rational structure.[44] This is why they assumed that for every issue of law they might encounter in the courts a right answer did exist, and it could be found (eventually) by men learned in the law. Tudor England's common lawyers often spoke of 'principles', 'rules', or 'maxims' when expressing their belief about the rational character of their law. During a lawsuit in 1551, a Serjeant of Law argued:

> There are two principal things from whence arguments may be drawn, that is to say, our maxims, and reason which is the mother of all laws. But maxims are the foundations of the law, and the conclusions of reason, and therefore they ought not to be impugned, but always to be admitted.[45]

A Tudor lawyer's 'maxims' were strictly limited practical directives. They included: 'inheritances may lineally descend but not ascend'; 'of every land there is a fee simple in somebody'; 'a dyscent taketh awaye an entre'; and 'the remainder ought to pass out of the lessor at the time of the livery'.[46] In 1549, Chief Justice Montague declared: 'But he that pleadeth an abatement of the writ, or pleads a plea after the last continuance, ought to plead it certainly; and these are observed as principles in our law.'[47] Tudor maxims were 'definite detailed legal rules of narrow content'.[48] They were exclusive professional tools, useful mnemonics for the practitioner at Westminster. 'Maxymes be onlye knowen in the kynges courtes or amonge them that take great studye in the lawe of the realme'; no layman could either arrive at or wield them.[49]

Yet 'maxims' were not seen by Tudor lawyers as statements of general 'principles', as the term is usually understood: that is, as initial premises or overarching structures (in this case, within the law). The common lawyers employed the terms 'principle', 'maxim' and 'rule' interchangeably, and it is precisely because they did not share the epistemological assumptions and the language of philosophers (and civil lawyers) that they could do this. 'That which our author here and in other places calleth a Maxime, hereafter he calleth a Principle; and it is all one with a Rule, a Common Ground, Postulatum, or an Axiome, and it were too much curiositie to make nice distinctions betweene them.'[50] Contemporary philosophers and Roman lawyers were dismayed by English lawyers' apparent abuse of what were for them highly technical terms. The effect was to reinforce the prevalent Continental opinion that the common law was barbarous and the legal learning of the English lawyers something less than a *scientia*.

Whatever the opinion of philosophers or civil lawyers, common lawyers continued to speak about their 'maxims'. They occupied a privileged place in the 'common erudition' and carried great weight in arguments among the lawyers: so much so that a common lawyer could assert 'they be of the same strength & effect in law as statutis be'.[51] Just like statutes, maxims 'were not to be questioned; they may only be interpreted and applied ... But they may be compared, apparent conflicts may be reconciled, and they may be justified by reason.'[52] (Any conflicts of maxims were 'apparent' and not real, since the law was a rational structure.) The 'Student of the Law' in Christopher St Germain's *Dialogue* (1532) declared there were six 'grounds of the lawe of England': the law of God, of reason, 'dyvers general customs of ... all the realme', the customs of particular places, statutes made in parliament and

> dyvers pryncyples, that be called by those learned in the lawe maxymes, the which have ben alwayes taken for lawe in this realme so that it is not lawfull for none that is lernyd to denye them; for every one of those maxymes is suffycyent auctorytie to hym selfe to such an extent that it is fruitless to argue with those who deny them. And whiche is a maxyme & whiche not shall alway be determyned by the Juges.[53]

The Tudor lawyers regarded their maxims as 'grounds' of the law, an attitude which stemmed from the fact that they were well-worn directives for the practitioner, 'givens' in their education at the Inns. We know they were derived from the professionals' activity of arguing and deciding upon points of law, but it is unlikely the Tudor lawyers ever chose to consider where their maxims originally had come from, or how they had become known to their predecessors, whose recital of them had been captured in the law reports they pored over as apprentices. To question them was to criticise the nature of their education and professional learning, and should a lawyer be thoughtless enough to do this he would cause great offence. By the sixteenth century, the maxims of the common lawyers were justified by professional faith alone. For all their conviction that their law was rational and all-embracing, the common lawyers placed strict limitations on how far one should question knowledge won from the 'hidden book' of the law. 'A maxime is a proposition to bee of all men confessed & granted without proofe, argument, or discourse. Contra negantem principia non est disputandum.'[54]

It was only because of a lawsuit, and a court trial, and only in the solemn setting of the court room that the common law could be discovered by the lawyers. Moreover, these revelations would be given only to sages of the law, men especially chosen and privileged by the monarch. It was the judges and the serjeants who were 'within the bar' of the court, and it was they alone who conducted the rituals of argument and public decision about what the law 'said'. The common law offered new pearls of wisdom only after formal rites and only to its temple priests, who then dispensed their new insights to the advocates of petitioners.

By the last years of Queen Elizabeth's reign, we can find new assumptions existing alongside (if not actually supplanting) the older ones about how to discover the hidden law. These assumptions, taken together, are about authority and its sources.

Legal printing had been an utterly practical innovation, but it eventually caused profound changes in the way lawyers studied and thought about the common law. Printed 'Year Books' gave the impression of comprehensiveness; indeed, they were bulky enough as to be unwieldy in use. As their manuscript collections of reports grew over their lifetime, lawyers had created so-called 'abridgements' to them which provided access by subject to cases. When the 'Year Books' appeared in print, so too did 'Abridgements', the most popular being Sir Anthony Fitzherbert's *Graunde Abridgement* of 1514 (often reprinted) and that of Sir Robert Brooke in 1573. 'These two great abridgements relieved the less industrious lawyers from having to make their own, and made it unnecessary for law students to read all the year-books through.'[55] From this shortcut flowed two perfectly natural but unexpected consequences. First of all, the subject divisions of the printed abridgements became canonical; these authors' groupings of cases, which every lawyer previously had invented for himself as an index to his own collection, were retained in the minds of Elizabethan law students as categories which really existed in the law; for example, 'contract', 'dette', 'enquest', 'examination', 'jurisdiction', 'the prerogative', 'obligacion', 'trespas', 'triall' and 'verdict'.[56] Secondly, the abridgement became accepted as an authority equal to the Year Book report itself, and in one instance at least as a greater authority: '[the case] was abridged by Fitzherbert more fully for the purpose here, than the [year] book at large in print is'.[57]

A similar process of attributing profound authority to texts was at work with regard to Sir Thomas Littleton's *Tenures* (*c.* 1481). 'Littleton', as his book on real property was familiarly known, had run through more printed editions by 1550 than that greatest source of law – the Bible.[58] Sir Thomas had represented it accurately as an introductory guide for his novice lawyer son, and it was recognised as an excellent elementary text, but the late Elizabethan lawyers regarded *Tenures* as such an authoritative source of the law, that 'Littleton' *was* the Bible on real property. When Sir Edward Coke published his own testament on the subject in 1628, he argued that 'Littleton' was not merely an 'ornament of the Common Laws' but 'the most perfect and absolute work that ever was written in any human science'.[59] Still more breath-taking was Coke's sub-title: *A Commentarie on Littleton, not the name of a lawyer onely, but of the Law it selfe.*[60]

This new willingness to concede authority to the printed page can be seen with 'Bracton' as well as with 'Littleton'. 'Bracton' was printed in 1569 for the first time, and in an expensive volume (the printer obviously knew he had a market).[61] After this, references and quotations from 'Bracton' can be found in several Elizabethan law treatises.[62] Lord Keeper (later Lord Chancellor) Thomas Egerton often cited 'Bracton' during his arguments.[63] Most importantly, in his *Reports* and in *A Commentarie on Littleton*, Coke cited 'Bracton' repeatedly. 'In Coke's writings Bracton was treated as an authority, that is, an author whose statements of law deserved acceptance unless contradicted by subsequent statute or judicial modification of the common law ... after Coke the denials and doubts as to Bracton's "authority" seem to fade.'[64]

Yet 'perhaps the deepest of all the changes in the Tudor period' was a change in what lawyers and litigants expected from those who judged their suits: men increasingly expected 'judgements with reasons'.[65] Plowden was bold enough to ask for the reasons for the decision in a case of 1570: 'And the justices did not argue the matter openly, but at last they agreed among themselves ... And because they did not openly declare the reasons of their judgement, I afterwards enquired them of Wray, Chief Justice of England.'[66] As a result of these new expectations, judges provided more elaborate justifications for their decisions, and the Elizabethan reporters naturally included details of these in their notebooks. The judges' decisions themselves were becoming an expected part of satisfactory law reports. 'Report ... in the common law it signifieth a publicke relation, or a bringing againe to memory cases judicially argued, debated, resolved, or adjudged in any of the king's courts of justice, *together with* such causes and reasons as were delivered by the judges of the same.'[67] These decisions became an important focus of the student's attention: younger lawyers now saw them as authoritative statements about the hidden law. Edmund Plowden, whose published reports were greatly respected by Elizabethan lawyers, boasted that

> other Reports be made most of the suddaine speech of the Judges upon motion of cases of the Serjeants and Counsellors at the Barre: But all the cases heere be matter in law tried upon Demurrers, or be special verdicts containing matters in law, of which the Judges had copies, studied them, and in most of them argued, and after great deliberation have given Judgement. And so (as I thinke) there is most firmnesse and suretie of Law in this Report.[68]

As the assumption spread that judgements should express the 'firmnesse and suretie of Law', so too did the thought that present judicial decisions should adhere to older ones, increasingly captured in Tudor law reports. 'Our booke cases', declared Sir Edward Coke, 'are the best proofes what the law is, *Argumentum ab auctoritate est fortissimum in lege*.'[69] Precedents now were gaining an authority they never had before, and 'precedent' lay not simply and generally in reports of old cases, but lay specifically in those judicial decisions which were made upon them. Coke emphasised this when he asserted 'that which of right is due to every reporter',

> that is, to reduce the sum and effect of [Judges' arguments] to such a method, as, upon consideration had of all the arguments, the reporter himself thinketh to be fittest and clearest for the right understanding of the true reasons and causes of the judgement and resolution of the case in question.[70]

Lawyers now began to work out how to use precedents in their arguments, and assigned differing weights of authoritativeness to different sorts of precedent and arguments from them. A single precedent had insufficient authority to convince the judges.[71] Two or three could not prevail against many precedents opposed to them.[72] If all the judges agreed in a decision, this became a very powerful precedent in its own right.[73] Parliamentary statutes were the most authoritative precedents of all,[74] but a concurrence of all the precedents *and* judges' opinions was almost as compelling.[75] Old precedents were not as weighty as recent ones.[76] Inferior courts must accept the customs of superior ones, but the reverse was not true.[77] A hierarchy of authority for precedents was rapidly being constructed by the professionals at Westminster. This structure reflected core assumptions of their guild, namely, the primacy of the present 'common erudition' and the superior authority of the judges and serjeants of the great central courts. Yet the very idea of precedents that compelled was a novel one, and it could take root only after the expectation of 'judgements with reasons' and the acceptance of printed texts as authoritative sources of the law had become unquestioned assumptions among (at least) some of the common lawyers.

It was in this late Elizabethan–early Jacobean context of new assumptions about the sources of authority that we find the beginnings of what in later centuries would become known as the 'case-law' method, or the 'logic of legal reasoning'. Put most simply, the term 'case-law reasoning' describes how a lawyer prepares his brief for a particular case and, to a considerable extent, it also decribes how he thinks about his trade. Common lawyers such as Coke and Bacon, whose assumptions about the law included the new ones discussed here, argued in the courts in this manner. The procedure can be outlined as follows: the present case is seen to be similar in important respects to older cases; a statement is fashioned which purports to encapsulate the essence of the judgements in those cases; and then the judge is asked to accept that the similarity of cases does indeed exist, and to deliver a similar judgement.[78] Unsurprisingly, the details are more subtle, especially since this outline would serve to describe the pattern of argument in 1450, as well as in 1600, 1800 and in the present. What had been added by the end of Elizabeth's reign was two things: first, the expectation that the law reports of cases would provide the inquiring lawyer with not only the judgement in a case but with the judge's reasons for it; and secondly, a greatly heightened respect for precedent. As a consequence of these new assumptions, the truly important similarities between the old cases and the present one were believed to be discovered by means of, yet behind, the reasons stated for those decisions (the *rationes decidendi*) and, in addition, present judges were expected to be swayed by arguments from precedent (even by abstract arguments):

> If the said imaginative rule be rightly and legally understood, it may stand for truth: for if you intend *ratio* for the legal and profound reason of such as by diligent study and long experience and observations are so learned in the laws of this realm, as out of the reason of the case they can rule the case in question, in that sense the said rule is true.[79]

What the lawyer searched for behind the *rationes decidendi* was an essential principle of law which those judges had come to know, and which had governed their particular decisions.[80] The lawyer would attempt to express this principle as best he could, and request its application to his client's case. This process can be illustrated by means of a chart:

1. Search law reports & abridgements for cases superficially similar to yours	case A	case B	case C	case D	case E
2. Find their *rationes*	(a)	(b)	(c)	(d)	(e)
3. Discover the principles of law underlying these	z	x	x	w? x?	x
4. The cases truly similar are		case B	case C	case D?	case E

(Case D is an ambiguous precedent, so a weak one to rely upon)

5. Then argue: (i) that your present case (P) is truly similar to cases B, C, E (& D); (ii) the judgements in cases B, C, E (& D) were based on the principle of law x; (iii) case (P), too, is governed by principle x; (iv) judgement in case P must be in accord with principle x, and thus in accord with that given in cases B, C, E, (& D).

This new way of arguing from old cases to the present one considerably reinforced all the common assumptions about the law being a substantial entity, rational in its hidden structure. Yet the procedure was a highly optimistic one with far-reaching implications. It required that hidden truths of the law be uncovered by the diligence of men learned in the law (of course, how any individual judge expressed them might well vary), and it assumed that these insights actually had governed the decisions of judges. Consequently, one could find sufficient licence here to begin thinking and speaking in terms of the 'principles' of the common law in much the same way that philosophers thought and spoke about the 'principles' of a learned *scientia*. Coke did so: 'It appeareth, that *jurisprudentia legis communis Angliae est scientia socialis* ... sociable, in that it agreeth with the principles and rules of other excellent sciences, divine and human.'[81]

We may note in passing just how the pattern of legal reasoning might be described by a logician. Persuasive claims have been made that the essential process is an inductive one,[82] that it is deductive,[83] that it is one of reasoning by analogy,[84] and that it is syllogistic.[85] Yet each of these claims isolates and privileges particular aspects of a lengthy process. It better reflects actual practice to view legal reasoning

> to be partly deductive and partly inductive, partly reasoning by analogy and partly the product of intuition, emotion or prejudice. In fact, legal reasoning will range over all the possible forms of human argument.[86]

Even an account as brief as this amply illustrates how sophisticated were the procedural and intellectual structures which typified the Elizabethan common law. Trials proceeded through formal and protracted stages: examining witnesses, establishing facts, reaching an issue, discovering the law, and delivering judgement. Complex professional beliefs existed about the nature of the law, and how (and which) sources of knowledge could be used to discover the right answer to a present dispute. This elaborate system had grown from modest beginnings in medieval kings' vigorous initiatives to spread royal justice and royal administrative authority throughout their kingdom. By Queen Elizabeth's reign, much of this royal initiative over shaping the courts and the justice they offered had been accumulated by the body of royal servants who worked in them, whose trade knowledge about working the machinery had developed quite naturally into a jealously guarded monopoly over the common law.

That Francis Bacon, like his father before him, was an excellent common lawyer is an observation readily accepted. But the observation disguises more than it illuminates. Sir Nicholas and his son *were* excellent common lawyers, and their contemporaries said so, yet both men were much more than lawyers. For the Bacons, father and son, 'lawyering' was a means to an end, not a career in itself. Mastering the law was to master the procedures and minutiae of effective governance. Practising at Westminster Hall and an apprenticeship in the royal administration was for them the best route by which to become a statesman.

Francis Bacon had climbed rapidly within Gray's Inn during the later 1580s: utter barrister (when aged twenty-one), bencher (aged twenty-six), and reader (aged twenty-seven). Many barristers never became benchers of their Inns; Bacon rose to bencher in fewer years than many men took to become utter barristers. Only his legal talent, and talent recognised and acknowledged, can account for these professional advancements within Gray's; these cannot be explained away simply as nostalgic favouritism toward the son of an illustrious (but dead) Lord Keeper.

From the mid-1580s and throughout the remainder of the queen's reign, Bacon was employed by Privy Councillors as a legal adviser, as an examiner of prisoners and their testimonies, and as a government propagandist. He is known to have argued in only four Elizabethan lawsuits, all of which were before the royal judges for discussion of fundamental issues of law: namely, those concerning the inheritance and conveyance of land, and those concerning contracts and the procedures for suits about them. Legal uncertainties about the security of land tenures were the most pressing of the two problems; they affected every estate in the kingdom, necessarily affected the Crown, and threatened the social order in general. Bacon seems to have made no attempts to build up a legal practice in the ordinary sense: his interests lay elsewhere.[87]

In the early 1590s (as we saw), Bacon was occupied with overseeing 'intelligence' on behalf of the Earl of Essex. The intense rivalry which grew up between Essex and Robert Cecil threw Privy Councillors, courtiers and officials into a turmoil: every vacant post, and every available patent and gift became a new site for skirmishes between the factions. Despite being partially compromised by his associations with the eventual loser in this power struggle, Bacon continued to enjoy the confidence of the queen and her councillors and he continued to be employed by them in highly sensitive matters.

This last point deserves wider attention, for it has been taken as certain that Bacon was disgraced in early 1593 for his criticism of a subsidy bill then before parliament, that the queen was deeply resentful, and that she thereafter blocked any senior appointments for Bacon. This helps explain why (or so the story goes) Essex, despite the expenditure of much political capital with the queen, was unable to obtain for Bacon the office of Attorney-General in 1594 or the Solicitor's post in the following year.[88] The queen was indeed upset by Bacon – less perhaps for making his objections in the first place than for failing subsequently to concede that he had erred – yet it does not follow that it was the business over the subsidy which kept him from becoming one of Elizabeth's law officers.

From the early summer of 1593, Essex (now a Privy Councillor) was pressing hard to win the office of Attorney-General for Bacon; in April of 1594 the campaign ended in embarrassing failure. Rather than simple pique about the subsidy episode (from both Bacon and the queen), in the final analysis it is more likely to have been Bacon's youthfulness and relative lack of experience that cost him the Attorney's place. No amount of blustering from Essex could have overcome this. Elizabeth's central government did not contain a great number of senior posts, and by the 1590s there existed to fill any vacancies a surfeit of ambitious and experienced officials in the middle ranks of the government machinery (not to mention the growing swell of ambitious but under-experienced gentlemen about the royal court and the Inns). Elizabeth's senior officials typically were appointed during their forties or fifties and after serving an arduous apprenticeship. For example, Thomas Egerton had been Solicitor since 1581 before his elevation to Attorney in 1592. Edward Coke, who followed Egerton as Attorney in 1594 at the age of forty-two, had been Recorder of Norwich (from 1586), Recorder of London (1591), the Queen's Solicitor since 1592 and Speaker of the Commons in 1593. Thomas Fleming was a Serjeant at Law and forty-one when finally he won Coke's place as Solicitor. It seems clear that maturity and experience in government were expected in a prospective law officer and, in particular, the office of Attorney-General was awarded only to lawyers who had proved themselves already in government administration.

At any rate, the queen did have confidence in Bacon's loyalty and his abilities, and she did not block Bacon from further legal appointments. At the very beginning of 1594, Bacon argued on three occasions in King's Bench and in Exchequer Chamber. He made a great impression on the judges and his audience; Burghley went so far as to write to congratulate him.[89] For six months now, Essex and Anthony Bacon had been doggedly investigating a circle of Spanish and Portuguese plotters and they eventually had uncovered a conspiracy to poison the queen through her private physician, Lopez, a Portuguese. Although he had had very little (or nothing) to do with these investigations, it was Francis Bacon who was commissioned in March 1594 to write the government's public advertisement presenting the details of the 'Lopez conspiracy' and pointing blame directly at King Philip of Spain.[90] In April, Bacon took Coke's place during 'Chudleigh's Case' in King's Bench, where he made another memorable argument. At about this time, he was appointed Deputy Chief Steward of the Duchy of Lancaster: a lawyer's post devoted largely to adjudicating land and property disputes arising within the Crown's Lancastrian domains.[91] In June, he was brought in to help the investigation into the 'Walpole plot'.[92] He was busy during August and September examining prisoners in yet another Catholic conspiracy.[93] Then, during Michaelmas Term 1594, the queen raised Bacon to the rank of 'Queen's Counsel Extraordinary', the first such appointment ever made. This brought Bacon 'within the bar' with the judges and serjeants at law and, in practical terms, it gave him a standing approximately that of the serjeants.[94] That Bacon should be brought within the bar and entitled 'Queen's Counsel Extraordinary' was not only a sign of favour and a reward for services rendered, but it was a precise acknowledgement of what his role actually was. Moreover, if it can be accepted that the queen was generally favourable to Bacon's suit for the solicitorship, to raise him first to the dignity of the serjeants at law was entirely appropriate. In short, Elizabeth's good will towards the young Bacon hardly could have been made more obvious.

In spite of his junior standing, it is clear that during 1594 and 1595 Bacon was favoured strongly for the solicitorship (vacant since

April 1594, when Coke became Attorney). Bacon was encouraged to function as the Solicitor, assisting Attorney Coke on several occasions in 1594 and 1595.[95] Many Privy Councillors (beside Essex these included Burghley, Cecil and Lord Keeper Puckering) spoke to the queen on Bacon's behalf. No other candidate seems to have been seriously considered for long. Yet Elizabeth's characteristic determination to maintain her autonomy and to be seen as withstanding pressure from her Privy Councillors resulted in delay after delay in announcing an appointment. By 1595, Bacon's minor indiscretions over the Subsidy Bill of 1593 were merely a convenient excuse for the queen's delays. And finally, it was Essex's ceaseless and public demands for his friend's appointment that were principally to blame for hardening the queen against it.[96] While Essex's interventions on his behalf became increasingly strident in 1595, Bacon's letters show him embarrassed and frustrated by the entire affair, for example:

> Against me [the Queen] is never peremptory but to my Lord of Essex ... My conceit is, that I am the least part of my own matter. But her Majesty would have a delay, and yet would not bear it herself. Therefore she giveth no way to me, and she perceiveth her counsel giveth no way to others, and so it sticketh as she would have it.[97]

In October 1595, writing to Lord Keeper Puckering, Bacon made abundantly clear not only his dissatisfaction but his distaste for Essex's factional struggles:

> his Lordship [Essex] may go on in his affection, which nevertheless myself have desired him to limit. But I assure your Lordship, I can take no furder care for the matter. I am now at Twicknam Park, where I think to stay ... though whensoever her Majesty shall like to employ me in any particular, I shall be ready to do her willing service.[98]

When Fleming was chosen for Solicitor in November 1595 instead of Bacon, Essex insisted on giving Bacon a parcel of land by way of compensation; Bacon immediately declared an unwillingness to be bound to him:

> My Lord, I see I must be your homager, and hold land of your gift; but do you know the manner of doing homage in law? always it is with a saving of his faith to the King and his other Lords; and

> therefore, my Lord ... I can be no more yours than I was, and it must be with the ancient savings: and if I grow to be a rich man, you will give me leave to give it back to some of your unrewarded followers.[99]

During the following year Francis Bacon drifted away from his formerly close association with Essex. One perhaps can understand why: on the one hand, it had become a liability, hindering his own chances of advancement in government and, on the other, it was becoming clearer that his counsels were not heeded by Essex. By 4 October 1596, Bacon could write bluntly to Essex, warning him of the serious political dangers inherent in his present course of action. As Bacon began by emphasising that his advice had been offered before, the letter is clearly a reproach as well. The assessment was not only shrewd but abrupt:

> How is it now? A man of a nature not to be ruled; that hath the advantage of my affection, and knoweth it; of an estate not grounded to his greatness; of a popular reputation; of a military dependence: I demand whether there can be a more dangerous image than this represented to any monarch living, much more to a lady, and of her Majesty's apprehension? And is it not more evident than demonstration itself, that whilst this impression continueth in her Majesty's breast, you can find [in courtiers] no other condition than inventions to keep your estate bare and low; crossing and disgracing your actions; extenuating and blasting of your merit; carping with contempt at your nature and fashions; breeding, nourishing, and fortifying such instruments as are most factious against you.[100]

It is possible that it was precisely because he withdrew himself from the infighting at court that Bacon's standing with the queen and her Privy Councillors remained high throughout the remainder of the reign. Bacon's position remained difficult. 'Queen's Counsel Extraordinary' was a formal post – of sorts – but it was without a pension or a regular means for accumulating fees; it was much more like personal service to the queen, and it depended fundamentally upon her continued trust in him. This meant avoiding being closely allied to any of the factions manœuvering at the court. In return for his balancing-act, Bacon enjoyed private access to the queen, which was among the most sought-after privileges at the royal court and an unusual honour for a man without senior rank or a principal office of state.[101] The number of Privy Councillors had averaged sixteen during the 1580s; in the 1590s Elizabeth allowed it to shrink to an average of twelve. When councillors whom she had long known and trusted died, the aging queen did not replace them; yet she created for Francis Bacon a special position from which he could offer her counsel.

The Privy Councillors, meanwhile, continued to employ Bacon – usually in harness with Coke, Fleming and William Waad (a senior clerk of the council) – for legal advice and especially for examining prisoners.[102] Moreover, in 1596 the Privy Councillors gave to Bacon the first of several special commissions. Bacon had been examining witnesses and prisoners at the behest of councillors for nearly a decade; what made this commission special (and those which followed in 1597 and 1598) was its explicit warrant to torture.[103] Commissions for the torture of prisoners were extremely rare in early modern England – later confusions about the Star Chamber process notwithstanding – but of the seventy-eight warrants we know about for the period 1540–1640, fifty-five were Elizabethan warrants and no fewer than forty of these were issued between 1586 (the beginning of the Spanish war) and 1598 (when Philip II died).[104] These were years of extreme anxiety by central governors about security, treachery and riot: so much so, that in addition to the 'ordinary' work of 'intelligence' and the interrogation of prisoners, Privy Councillors were prepared now to deploy the extraordinary weapon of torture several times a year in order to root out plots and conspirators.[105] Given such a climate of apprehension, the four commissions addressed to Bacon underline yet again the councillors' great confidence in him. The first of these was directed to 'Mr Attorney and Mr Sollycitor General, Mr Francis Bacon, and Mr Recorder of London, *or to anie two of them*'.[106] The second warrant named Coke, Fleming, Bacon and Waad, and the next two were for Fleming, Bacon and Waad alone. Such warrants usually were addressed to the commissioners in order of precedence; when taken together, they clearly reflect what the Privy Council understood to be Bacon's rank and function: it was high (adjacent to that of the Solicitor) and intimately associated with the security of the queen and the regime.

Not surprisingly, Bacon was included in several committees of the parliaments of 1593 and 1597 drafting security and policing legislation.[107] He appeared before the Privy Councillors in the Star

Chamber in 1596 and 1597, both as Crown counsel and as an examiner.[108] Equally related to his security work was Bacon's repeated employment by councillors as a government propagandist, justifying the regime's religious policies and publicising the Crown's case against conspirators in such a way as to sustain both public hatred of the papacy and Philip of Spain, and public support for the regime's foreign policies. In 1589 he wrote the 'Advertisement' about the Church of England, and a statement of English religious and foreign policy known as 'Sir Francis Walsingham, Secretary, to Monsieur Critoy, Secretary of France'. In 1593 he responded to the attack upon the queen and English policies made by Robert Parsons, the Catholic propagandist, by writing *Certain Observations made upon a Libel.* A year later appeared *A True Report of the Detestable Treason Intended by Dr Roderigo Lopez*, and in 1599 *A True Report of a Strange Conspiracy*, the story of the Squire and Walpole plot against the queen.[109]

Given his multi-faceted duties as 'Queen's Counsel Extraordinary', his expressed reservations about Essex and his overriding concern for the security of the queen and his allegiance to her regime, it was natural that Bacon was expected to participate in both the investigations and the trial of Essex for his rebellion in 1601. It was equally natural that Bacon should have participated so wholeheartedly in the earl's prosecution and thereafter written the detailed advertisements describing the conspiracy and justifying the Crown's actions. Essex's feeble accusation that Bacon had betrayed his friendship – an accusation adopted uncritically by the populace (and subsequently) – ignored the significance of Bacon's repeated advice and warnings, and particularly his warning about the nature of homage and fealty: his loyalties were first and foremost to the queen and the regime.[110]

We have argued that an appreciation of Bacon's own legal employments and of the workings of English law was necessary for understanding Bacon. Such a consideration is especially important for us since Bacon adopted as sound procedures in natural philosophy several procedures for 'discovery' used among common lawyers. By looking at his legal activities in the 1590s, we see that all these procedures were utterly familiar to him. Yet we have found much more as well. 'Law' was not simply a structure of rules and institutions for coercion and punishment; the term encompassed not only the ways by which disputes were settled, but the institutional structure of governance in the realm, and the intellectual world of the lawyers' guild as well. To become 'learned in the law' was simultaneously to become learned in the practical minutiae of governance in England. In addition, common lawyers shared assumptions about how 'to establish facts', how to argue, and about the nature of the law and how to uncover its hidden truths. Their training and practice provided Elizabethan common lawyers with distinctive beliefs about knowledge and how one successfully ascertained truths. Thus the common lawyers had a particular stance (albeit a largely implicit one) about what philosophers now choose to call 'epistemology', or the theory of knowledge. Francis Bacon was 'learned in the law'; he shared certain convictions with his fellow lawyers about 'The Law', and he was actively employed in government as well. Yet his perspective on governance was not that of a private barrister nor that of a subordinate official, a mere administrator. He saw himself as a councillor and statesman, and he viewed the nation and the machinery of governance from the elevated perspective of an 'extraordinary' personal councillor to the queen. In particular, he was deeply anxious about the security and stability of the regime: he worried about assassination, and about anything that might provoke riot, rebellion or disorder in the state. As a consequence, Bacon had firmly authoritarian ideas about how the state should be organised and governed. These, too, find their place in his reformed natural philosophy, as we shall see.

5

A reformed state

> All those which have written of laws, have written either as philosophers or as lawyers, and none as statesmen. As for the philosophers, they make imaginary laws for imaginary commonwealths; and their discourses are as the stars, which give little light because they are so high. For the lawyers, they write according to the states where they live what is received law, and not what ought to be law: for the wisdom of a lawmaker is one, and of a lawyer is another.
>
> *Advancement of Learning* (1605), *Works*, III, p. 475

During the 1590s and throughout the remainder of his life, Bacon argued for legal reform. Just like the phrase itself, campaigns for 'legal reform' can mean very different things: their proponents can be hoping for changes in specific laws, or for changes in the administration of justice and governance, and even for changes in what is understood by 'The Law'. For Bacon, legal reform meant change of all three sorts, and for a very specific end.[1]

As to the first sort of legal reform, changes in particular laws, one could argue that Bacon was hardly unique. As they have been whenever lawyers and men of affairs have congregated, such considerations were staple fare in lawyers' conversation in Elizabethan times, and Members of Parliament were routinely concerned with repealing, amending or drafting legislation. Similarly, the second sort of reforms were of interest to local magistrates, and (for obvious reasons) they were of especial concern to the Privy Councillors, the Royal Justices and the other legal officers of the Crown. The exhortations of Lord Chancellors to successive parliaments about the need for disciplined, uniform and equitable exercise of justice throughout the country is a clear example of the Privy Councillors' attention to the practical difficulties of governance. Yet certainly, Bacon's particular proposals of legal reform go far beyond what could be expected of a lawyer, an MP, a

JP, and one of the government's legal advisors. By 1596 his views about the nature of 'The Law', and how one should make discoveries in it, differed greatly from those voiced by contemporary common lawyers. In *The Maxims of the Law*, which he presented to the queen during the Christmas season of that year, Bacon revealed highly unorthodox opinions about the nature of 'The Law'. When, with this in mind, we reflect upon his other pronouncements, we find they are all of a piece, and with them we can reconstruct Bacon's grand programme of legal reform. Doing so is crucial, since it was Bacon's reformed law which provided the blueprint for a reformed natural philosophy.

Such a sweeping plan had profound implications for the structure of governance in the state as well. When Bacon spoke to King James about a 'Greater Britain', he was not thinking of a state which was 'greater' merely because the English and Scottish kingdoms had been combined. 'Greater Britain' was to be a state with a structure of governance such that it could become a prosperous empire. This imperial polity can be seen in 'Bensalem' as well, the island state of Bacon's utopian tale, *New Atlantis*, and – yet again – in the polity of Bensalem's most famous institution: Salomon's House, the home of researches into the secrets of nature.

Law reform

In Bacon's 'Speeches of the Six Counsellors', written for the Gray's Inn Christmas revels of 1594–5, the 'Fifth Counsellor' advised the pursuit of 'Virtue and a gracious Government', imploring his prince to

> look into the state of your laws and justice of your land; purge out multiplicity of laws, clear the incertainty of them, repeal those that are snaring, and press the execution of those that are wholesome and necessary; define the jurisdiction of your courts, repress all suits and vexations, all causeless delays and fraudulent shifts and devices, and reform all such abuses of right and justice; assist the ministers thereof, punish severely all extortions and exactions of officers, all corruptions in trials and sentences of judgement.[2]

Although it was offered on this occasion to the imaginary 'Prince of Purpoole', Bacon believed this advice and repeatedly offered it to Queen Elizabeth and King James. While attacking tirelessly the 'multiplicity' and 'incertainty' of the laws, Bacon pursued a campaign for law reform on four fronts: he proposed to 'purge' the statute books; he argued for a system of official legal reporting and better law reports; he wished to reinforce the judicial hierarchy, and to alter the role of the royal justices in the state; and he argued for a view of the nature of 'The Law' and for methods of discovering the hidden law quite at odds with those of other common lawyers. What was the substance of his campaign, and what did he expect to achieve by it?

Speeches which deplored the confusion produced by the tangle of penal laws were heard in every parliament of the 1590s. At the beginning of the 1593 parliament, Lord Keeper Puckering's plea to this effect was promptly brought to the attention of the Commons by Francis Bacon, who argued:

> The cause of the assembling of all parliaments hath been heretofore for Laws or Money: the one being the sinews of Peace, the other of War ... I did take contentment in her Majesty's speeches the other day delivered by the Lord Keeper ... to purge the statute book [and] lessen the volume of laws; being so many in number that neither common people can half practise them, nor the lawyer sufficiently understand them.[3]

In 1597, the new Lord Keeper (Thomas Egerton) made a plea similar to that voiced by his predecessor, and shortly after his speech a motion in the Commons called for 'the abridging and reforming of the excessive number of superfluous and burdensome penal laws'; it was seconded by Bacon, but nothing came of it.[4] In the 1601 parliament, Bacon argued: 'I know and do assure myself there are many more than I know laws both needless and dangerous.'[5] In making these remarks, Bacon might be thought to be doing little more than echoing the sentiments of the queen and her Lord Chancellors: there were too many laws and pruning was needed, and since other laws contained obscurities and ambiguities, better drafting was required. But what is striking in Bacon's speeches were his proposals for solving the problem. In 1593, he reminded the Commons:

> The Romans appointed 10 who were to correct and recall all former laws, and set forth their 12 Tables; the Athenians likewise appointed 6 to that purpose; Lewis XI of France did the like in reforming the laws.[6]

In 1601, he argued for a standing committee of the House 'for the repeal of divers statutes, and of divers superfluous branches of statutes . . . The more laws we make the more snares we lay to trap ourselves.'[7] In 1604, during the debates on the difficulties of a union between the two kingdoms, Bacon wrote to King James, proposing the preparation of a 'digest':

> For the Laws, to make an entire and perfect union, it is a matter of great difficulty and length, both in the collecting of them, and in the passing of them . . . as to the collecting of them, there must be made by the lawyers of either nation a digest, under titles, of their several laws and customs, as well common laws as statutes; that they may be collated and compared, and that the diversities may appear to be discerned of.[8]

Bacon's proposed solution, then, involved committees of experienced lawyers who would collect the laws in the statute books, arrange them under various headings in 'digests', and then collate and compare them. Thus the laws which were common to both countries could be identified, the conflict of laws discovered as well and new laws applicable to both countries could be drafted and brought before parliament. In these remarks we see Bacon concerned with eliminating redundancies and ambiguities within the statute book; to this end he proposed a bureaucratic solution: standing committees of experts sifting laws into a 'digest', and having also the mandate to scrutinise the laws and prepare a list of the acceptable laws of the land.

A conventional truth about English law, a truth which its practitioners repeatedly offered up, was that it included unwritten and 'customary' law as well as a statute book. Yet in writing to the king in 1614, Bacon argued that however much 'the Common Law of England, it appeareth it is no Text law', nonetheless,

> the substance of it consisteth in the series and succession of Judicial Acts from time to time which have been set down in the books which we term Year Books or Reports, so that, as these reports are more or less perfect, so the Law itself is more or less certain, and indeed better or worse.[9]

English common law for practical purposes was indeed a 'Text law'. By way of offering solution for the imperfections and uncertainties in the books, Bacon addressed himself both to the existing law reports, and to the better generation of future ones. The reports currently available should be 'purged and reviewed, whereby they may be reduced to fewer volumes and clearer resolutions'. Such a scrutiny could be achieved thus:

> By taking away many cases obsolete and of no use, keeping a remembrance of some few of them for antiquity sake.
>
> By taking away many cases that are merely but iterations, wherein a few set down will serve for many.
>
> By taking away idle Queries, which serve but for seeds of incertainty.
>
> By abridging and dilucidating cases tediously or darkly reported.
>
> By purging away cases erroneously reported and differing from the original verity of the Record.
>
> Whereby the Common Law of England will be reduced to a corse or digest of Books of competent volumes to be studied, and of a nature and content rectified in all points.
>
> Thus much for the time past.

The 'purging' techniques here are similar to those he proposed for the statute books.

Bacon went on to urge arrangements for future law reporting. He argued for the 'restoration' of official reporters to the royal courts at Westminster,

> which in former times were persons of great learning . . . and did carefully and faithfully receive the Rules and Judicial Resolutions given in the King's Courts, and had stipends of the Crown for the same.[10]

Moreover, the reports of these royal officers were to supplant those made 'of later times by the industry of voluntaries',

> as chiefly by the worthy endeavours of the Lord Dier and the Lord Coke. But great Judges are unfit persons to be reporters, for they have either too little leisure or too much authority, as may appear well by those two books, whereof that of my Lord Dier is but a kind of note book, and those of my Lord Coke hold too much *de proprio*.

In other words, Dyer's reports, and those of Coke, which were fast becoming the common lawyers' preferred body of modern law reports, were unacceptable and needed to be replaced. Dyer, too busy, had left little more than his notes and Coke, too assertive, had built up his reports with opinions peculiarly his own.

Closely related to his concerns about law reports were Bacon's proposals to examine the ancient records of government, and then compile a book *De antiquitatibus juris*:

> All ancient records in your Tower, all the acts of parliaments, letters patents, commissions, and judgements, and the like are to be searched, perused, and weighed. And out of these are to be selected those that are of most worth and weight; and in order of time, not of titles (for the more conformity with the Year Books) to be set down and registered; rarely in *haec verba*; but summed with judgement, not omitting any material part. These are to be used for reverend precedents, but not for binding authorities.[11]

Bacon had much to say as well about the role of judges under the king, and about the structure of justice in the state. In 1607, Bacon wrote 'A Proclamation for Jurors', which was published in the autumn of the year as a royal proclamation. He declared that the king wished

> to extend our princely care to another sort of judges (though they be termed by another name) upon whom lieth a principal part of judicature, which are the Judges of the fact ... which try and decide the issues and points of fact in all controversies and causes; – a matter no less important to the sum of justice than the true and judicious exposition of the laws themselves ... unto which sort of Judges also the law of our realm doth ascribe such trust and confidence, as it neither ties them to the evidence and proofs produced, neither disableth any witness (except in case of perjury) to be used; but leaveth both supply of testimony and the discerning and credit of testimony wholly to the Juries' consciences and understanding, yea to their private knowledge ... so nevertheless it is then laudable and good when those persons ... are men of such quality, credit, and understanding, as are worthy to be trusted with so great a charge as to try men's lives, good names, lands and goods, and whatsoever they hold dear in this world.[12]

Yet the proclamation was not simply laudatory. Bacon proposed a return to a state of affairs which supposedly had existed 'before abuses crept in', namely that all (and only) those men 'which have freehold according to the law ... shall be returned to serve upon Juries':

> Time and abuse have so embased the estimation of this service, and altered the use thereof, as sheriffs, under-sheriffs, and bailiffs do not only spare gentlemen of quality in a kind of awe and respect; but do likewise for lucre, gain, and reward, forbear to return many of the ablest and fittest persons; so that the service oftentimes resteth upon such as are either simple and ignorant, and almost at a gaze in any case of difficulty; or else so accustomed and inured to pass and serve upon Juries, as they have almost lost that tenderness of conscience which in such cases is to be wished, and make the service as it were an occupation or practice.

Not only were sheriffs allowing gentlemen to avoid their responsibilities, and taking rewards for doing so (thus compromising their royal office), but lesser men, who had no legal right to assume the burden, were serving as jurors. The proclamation sought to bolster the structure of governance in the land: the royal government, in the person of the sheriff, was to appoint jurors solely from the 'men of quality, credit and understanding' upon whom 'lieth a principal part of judicature'. It is sufficient for our immediate purpose to note that the proclamation asserted that gentlemen were not only to be the Justices of the Peace, but they were to be the juries as well.

In 1616, Bacon wrote a long and comprehensive letter laying down precepts for the management of matters of religion, justice, the Privy Council, foreign affairs, 'peace and war', trade, colonies, and the royal court itself.[13] Bacon was at pains to delimit the role of Parliament. Parliament, he argued, had a specific and highly limited place in the government of the realm: it was *not* one of the 'true organs' of governance. He affirmed the view that the 'true constitution' of the 'High Court of Parliament' consisted of the two Houses of Lords and Commons and 'the King's Majesty, as the head of that great body', but he argued that the two Houses were 'more properly a Council to the King, the great Council of the Kingdom, to advise his Majesty of those things of weight and difficulty which concern both the King and kingdom, than a court'.[14] Although it was true that 'no new laws can be made, or old ones abrogated and altered, but by common consent in Parliament', the reason Bacon offered for the distinction between a 'Council' and a 'court' was that in parliament no laws were 'concluded but by the King's royal assent, in person or by his commissioners delegated; they are but embryos, till he gives life unto them'. The king was the only lawgiver. Parliaments, he continued, were not to sit longer than absolutely necessary,

> for then they are but burthens to the people, by reason of the privileges due to the members of the two houses and their attendants; which their privileges and rights are religiously to be observed and preserved; but if they should be unjustly enlarged beyond the true bounds, they might lessen the just power of the Crown, it borders so near upon popularity.

In the parliament of 1610, Bacon had argued:

> If the King's sovereignty receive dimunition or any degree of contempt from us that are born under a hereditary monarchy (so as the motions of our estate cannot work in any other frame or engine) it must follow that we shall be a *meteor* or *corpus imperfecte mistum*; which kind of bodies come speedily to confusion and dissolution.[15]

Parliament was undoubtedly a sacred council, but it was to be carefully controlled lest it lead to a disastrous undermining of the royal powers.

Bacon's precepts about the judges and the institutions of royal judicature are highly revealing. In the first place, of course, justice was the king's alone, but he was represented by his judges: 'The honour of the Judges in judicature is the King's honour, whose person they represent.' Bacon argued that despite the assize judges' great assistance in hearing occasional private suits and in delivering the gaols,

> yet they are of more use for the government of the countries through which they pass, if that were more thought upon. For if they had instructions for that purpose, they might be the best intelligencers to the King of the state of the whole kingdom, of the disposition of the people, of their inclinations, of their mutations and motions, which are necessary to be understood truly.[16]

By the king (or by his Lord Chancellor 'in the King's name') the royal judges should be given a 'charge of those things which the present times did most require; and at their return should deliver a faithful account thereof, how they found and left the countries through which they passed'.

In May of 1617, upon his installation as Lord Keeper of the Great Seal, Bacon made another proposal:

> It falls out, that there be three of us the King's servants in great place, that are lawyers by descent, Mr. Attorney son of a Judge, Mr. Solicitor likewise son of a Judge, and myself, a Chancellor's son.
> Now because the law roots so well in my time, I will water it at the roots thus far, as beside these great ones, I will hear any Judge's son before a Serjeant, and any Serjeant's son before a Reader.[17]

Lawyers whose fathers had devoted themselves to the king's business of judicature were to be favoured. The privilege of prior audience before the Court of Chancery would bring practical benefits to a lawyer, since clients would flock to a man who could have suits heard promptly. And to dignify judges' and serjeants' sons in this manner was simultaneously to enhance the importance of royal judicial service as opposed to private legal practice. The effect of such positive discrimination was to elevate the sons of lawyers above other lawyers, and it was the result of Bacon's insistence that the judges and serjeants were among the premier officers in the king's government.

Bacon was lavish in his praise of the system of circuits. The circuits were travelling displays of the monarch's authority and of his great gift of justice to his people, and they also displayed the features Bacon wanted in the king's officials: loyalty to the Crown, independence from local interests, and active involvement in the royal governance of the realm. Before the judges went on their circuits in summer 1617, Bacon told them:

> You that are the Judges of Circuits are as it were the Planets of the Kingdom ... and no doubt you have as great stroke in the frame of this government, as the other have in the great frame of the world. Do therefore as they do; move always and be carried with the motion of your first mover, which is your Sovereign. A popular Judge is a deformed thing: and *plaudite's* are fitter for players than for magistrates. Do good to the people, love them and give them justice. But let it be, as the Psalm saith, *nihil inde expectantes*: looking for nothing, neither praise nor profit.[18]

On a similar occasion in February of 1618, he argued that 'the six Circuits of England are like the four rivers of Paradise, they go to water the whole kingdom, and pass through the whole land to the distributing of justice'.[19] These are striking images indeed, and Bacon was serious in using them: in his eyes, the royal judges possessed an awesome responsibility and should hold a principal place in the system of royal governance.

Writing to King James in 1616, Bacon offered a 'way to reduce and recompile the Laws of England':

> This work is to be done ... in this manner. It consisteth of two parts; the digest or recompiling of the common laws, and that of the statutes.[20]

We now have heard many of Bacon's thoughts about reforming the statute book, 'pruning' the law reports, enhancing the role of the judges, and bolstering the judicial hierarchy. In this letter, Bacon touches on all the elements in his legal reform programme, and we can begin to see the complete picture.

As far as the statute book was concerned, it seems not only that the king had listened to Bacon in years past, but that he had assented to his forming a committee for 'the reducing of concurrent statutes, heaped one upon another, to one clear and uniform law'.[21] Bacon's opinions about what to do with the statutes remained as they had been when he first expressed a desire to 'recompile' them in the 1590s. But he now gave much specific information about the other 'part' of the reform programme – the 'recompiling of the common laws':

> Three things are to be done:
> 1. The compiling of a book *De antiquitatibus juris*.
> 2. The reducing or perfecting of the course or corps of the common laws.
> 3. The composing of certain introductive and auxiliary books touching the study of the laws.

As for the second point, which dealt with the 'corps of the common laws' presently available, Bacon desired there 'to be made a perfect course of the law *in serie temporis*, or Year Books (as we call them) from Edward the First to this day', a 'perfect course' which he had discussed often in years past. Regarding a work *De antiquitatibus juris*, Bacon was here advancing yet another claim: that the ancient records of royal governance ('acts of parliament, letters patents, commissions, and judgements, and the like') were not to be regarded as current law. Selections from these records would form the public text *De antiquitatibus juris*, and 'these were to be used for reverend precedents, but not for binding authorities'. Bacon's third 'thing to be done' brings us closer to a positive understanding of just what Bacon envisioned as the proper science of the law:

> For the auxiliary books that conduce to the study and science of the law, they are three: Institutions, a treatise *De regulis juris*; and a better book *De verborum significationibus*, or terms of the law.

The book *De significationibus* was to contain 'not only the exposition of the terms of the law, but ... the words of all ancient records and precedents'. This could only be based on the new book *De antiquitatibus juris* and the revised law reports. As for 'Institutions', the 'office' of such a work was 'to be a key and general preparation to the reading of the course'.[22] Bacon accepted that Littleton and Fitzherbert had written good 'books of introductions' but these, he argued, 'are no ways of the nature of an *Institutions*'. What he wanted was a work with two properties: 'the one a perspicuous and clear order or method; and the other, a universal latitude or comprehension, that the students may have a little prenotion of every thing, like a model towards a great building'. But on what basis was such a work of 'prenotion' to be formed? To talk in such a way about how to study the 'corps of the common laws' was to imply beliefs about the nature of the law. If students were to be given a 'prenotion of every thing', then not only must the law itself be a stable entity, but it must have major structures in it which all men could grasp. The first belief was widespread among the common lawyers, but the second most emphatically was not. In Bacon's eyes, 'The Law' could be seen as a 'great building' with structures in it. Furthermore, his 'Institutions' was a primer written after many of the building's structures were known; only then could it serve the student as a 'model'. An 'Institutions' would teach a 'perspicuous method' as well; and this was the method behind the proposed treatise *De regulis juris*, which was a repository of known structures of the law, and would guide one's subsequent learning in the law.

'Rules', 'regulae', 'maxims' or 'grounds' were words all common lawyers knew as referring to certain hallowed practical directives. But Bacon dismissed these:

> For the treatise *de regulis juris*, I hold it of all other things the most important to the health, as I may term it, and good Institutions of

> any laws; it is indeed like the ballast of a ship, to keep all upright and stable; but I have seen little in this kind, either in our law or other laws, that satisfieth me. The naked rule or maxim doth not the effect. It must be made useful by good differences, ampliations, and limitations, warranted by good authorities; and this not by raising up of quotations and references, but by discourse and deducement in a just tractate.

Bacon had worked upon such a 'tractate', and he had placed very high hopes upon it:

> In this I have travelled myself, at the first more cursorily, since with more diligence, and will go on with it, if God and your Majesty will give me leave. And I do assure your Majesty, I am in good hope, that when Sir Edward Coke's Reports and my Rules and Decisons shall come to posterity, there will be (whatsoever is now thought) question who was the greater lawyer?

As long ago as the early 1590s, Bacon had begun just this sort of 'travell'. One version of it is all that remains of his labour on *regulae juris*, but the tract called *Maxims of the Law*, which he presented to Queen Elizabeth in late 1596, tells us a very great deal about what he thought the law to be, and the methods by which one could discover truths hidden within it.[23]

In dedicating *Maxims of the Law* to 'Her Sacred Majesty', Bacon made statements about the purpose of his work some of which are familiar to us now from the writings we have reviewed, and he claimed as well that it was a purpose 'for these many years infused in your Majesty's breast'

> to enter into a general amendment of the state of your laws, and to reduce them to more brevity and certainty; that ... the snaring penalties that lie upon many subjects [be] removed; the execution of many profitable laws revived; the judge better directed in his sentence; the counsellor better warranted in his counsel; the student eased in his reading ... Which purpose and intention, as it did strike me with great admiration when I heard it, so it must be acknowledged to be one of the most chosen works, of highest merit and beneficence towards the subject, that ever entered into the mind of any king: greater than we can imagine.

For some time now, 'by my private travayle', Bacon had been collecting 'many of the grounds of the common laws', and he had been doing so 'the better to establish and settle a certain sense of law which doth now too much waver in incertainty'. In the preface, he claimed he wished to 'visit and strengthen the roots and foundations of the science itself', and was doing so by a particular method: 'by collecting the rules and grounds dispersed throughout the body of the ... laws'. This is not what his contemporaries understood about the 'rules and grounds'; for them these were not 'dispersed throughout' the body of the laws at all, and therefore could not be extracted from it by study. But Bacon was convinced his approach was warranted,

> for here no small light will be given, in new causes and such wherein there is no direct authority, to sound into the true conceit of law by depth of reason; in cases where the authorities do square and vary, to settell the law, and to make it received one way; and in cases wherein the law is cleared by authority, yet nevertheless to see more profoundly into the reason of such judgements and ruled cases, and thereby make use of them for the decision of other cases more doubtful ... Neither will the use thereof be only in deciding of doubts, and helping soundness of judgement, but further, in gracing of argument; in correcting of unprofitable subtlety, and reducing the same to a more sound and substantial sense of law; in reclaiming vulgar errors, and generally in the amendment in some measure of the very nature and complexion of the whole law. And therefore the conclusions of reason of this kind are worthily and aptly called ... *legum leges*.

Bacon's *legum leges*, then, were to be the 'conclusions of reason', and the aim was nothing less than 'the amendment ... of the very nature and complexion of the whole law'. Because such *legum leges* would be of particular use to judges and pleaders in suits which turned on 'issues of law', whensoever Bacon's 'rules and grounds' were to be followed, then an 'amendment ... of the whole law' would indeed be taking place. All three of the situations which he addressed here were faced by judges in the courts. In novel causes ('where there is no direct authority'), the judge now would have to hand a set of reliable 'rules' or principles upon one of which a judgement could be based. Similarly, in cases in which 'the authorities do square [i.e., clash] and vary', he would be able to consult appropriate *legum leges* and thereby 'settell the law' so that thereafter it would be 'received one way'. If the law was fairly clear, he could better understand the 'conclusion of reason' behind relevant earlier judgements and be more confident about any future application or extension of that reasoning.

Bacon was at pains to explain his 'manner of setting down' his 'rules' and this again helps us understand what he was trying to do. Some of them, he thought, had a 'concurrence with the civil Roman law', but he was happy then to use the form of words that 'the civilians use' since he considered it 'a matter of greater authority and majesty, to see and consider the concordance between the laws penned and as it were dicted *verbatim* by the same reason'. Moreover, he thought it best to leave them in 'distinct and disjoined aphorisms'

> whereas I could have digested these rules into a certain method or order, which, I know, would have been more admired, as that which would have made every particular rule, through his coherence and relation unto other rules, seem more cunning and more deep; yet have I avoided to do so, because this delivering of knowledge in distinct and disjoined aphorisms doth leave the wit of man more free to turn and toss, and to make use of that which is so delivered to more several purposes and applications. For we see all the ancient wisdom and science was wont to be delivered in that form; as may be seen by the parables of Solomon, and by the aphorisms of Hippocrates, and the moral verses of Theognis and Phocylides: but chiefly the precedent of the civil law, which hath taken the same course with their rules, did confirm me in my opinion.

Bacon's *legum leges* would incorporate some of the 'maxims' currently used by the common lawyers. He chose not to neglect these, although they were 'ordinary and vulgar', but

> having chosen out of them such as doe yield profitable erudition, I have reduced them to a true application, limiting and defining their bounds, that they may not be runned upon at large, but restrained to point of difference. For as, both in the law and other sciences, the handling of questions by commonplace, without aim or application, is the weakest; so yet nevertheless many common principles and generalities are not to be contemned, if they be well derived and deduced into particulars, and their limits and exclusions duly assigned.[24]

That is to say, he was presenting them in his own words, not in the form of words in which they were commonly found.

Others among his rules were 'gathered and extracted out of the harmony and congruity of cases' found in the reformed law reports. And here (just as with the reform of the statute book), Bacon envisioned a process involving comparison, identification of the common parts, and 'extractions' (elaborations) built upon these 'congruities': the very process earlier described as the logic of legal reasoning.[25] Yet Bacon was proposing here to do more. He planned to offer the judges an official catalogue of the principles upon which their *rationes* and their judgements were to be based.

The several parts of Bacon's grand plan for legal reform can be displayed in charts. First of all, Bacon proposed to engage committees of experts to sift information and create 'digests':

Secondly, from the new law reports Bacon expected (by 'dint of reason') to generate *regulae juris*, derived from which was a students' book called 'Institutions'. The catalogue of *regulae* would help judges make their judgements upon suits which turned on 'issues of law'. These judgements would be incorporated into the body of law reports by the official law reporters and, in swelling the law reports, provide materials for further reflection and the generation of more *regulae*:

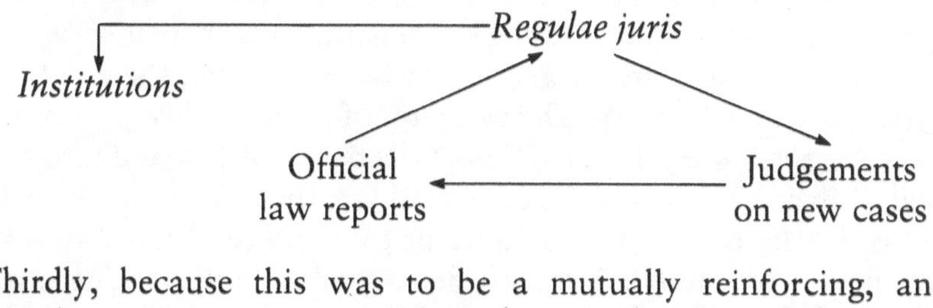

Thirdly, because this was to be a mutually reinforcing, and growing, system, we can represent the growth in knowledge in the following manner:

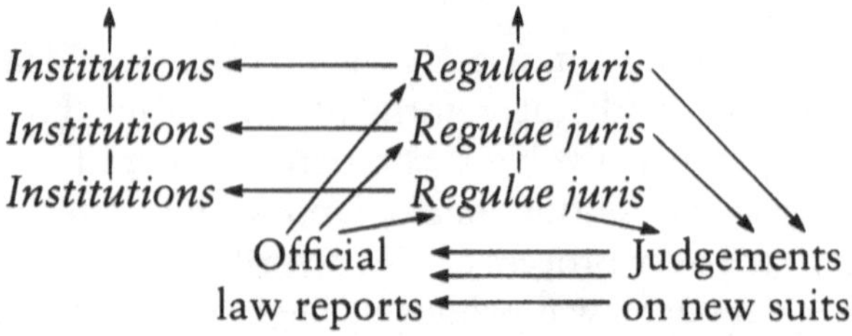

The law reports, therefore, would become more and more certain and useful, since new 'issues of law' increasingly would be decided with reference to *regulae juris*, and the judgements and *rationes* would become clearer, more explicit, and less idiosyncratic. Thus the discovery of new, or more refined, or carefully restricted *regulae* would be easier and faster, since in looking at the official reports one could more readily grasp the reasoning (and *regulae*) behind the judgements. By allowing *regulae* to stand in the first instance in the form of 'disjoined aphorisms', Bacon planned to guard against excessive rigidity in his *regulae*, since aphoristic expression is flexible enough to allow for the future refinements and splitting that his system was designed to achieve. As the process advanced, some of the early *regulae* would be 'split', providing stable, delimited, rules; others would remain in flexible aphoristic form, continuing to provoke new insights and connections. After all, this was the force of Bacon's references to the aphorisms of Hippocrates and the parables of Solomon: these continued to be potent food for scholarly thought.

Having described Bacon's proposals for legal reform, we now can consider their implications. Bacon's 'Proclamation for Jurors' of 1607 was designed to restore royal control over judicature, i.e., governance, in the provinces. Serious 'abuses had crept in'. The sheriffs were failing in their duty to the king to appoint only 'men of quality' to juries and, by bribing or pressuring the sheriffs to choose lesser men in their stead, gentlemen were avoiding their duty to participate in judicature. The selection of a sheriff, like that of the MPs 'for the Shire', the members of the bench of JPs (and jurors as well, if Bacon had his way), depended greatly upon a man's status in his district or county. The ancient royal office of the shrievalty was now typically filled in accordance with a local pecking order and local politics. In Bacon's eyes, the effect of recent 'abuses' was that the integrity of the king's system of justice was suffering. Jury duty was supposed to be a mark of honour, but Bacon thought it was being sneered at, and deteriorating into a mere 'occupation' of the simple and ignorant. He was hoping that royal service would become independent of local status and interests, that men who served the king would serve him alone:

> The gentlemen of best quality [must] put away that vain and untrue conceit that they are any ways disgraced or disesteemed, if they be called upon or used in this part of Justice to be Judges of the fact; knowing that all judgement is God's principally, and by him committed unto us within the precinct of our kingdoms as his minister upon earth, to whom likewise they are subordinate.[26]

The proclamation sought simultaneously to affirm a social hierarchy in which the authority and the duty of gentlemen to participate in the governance of their communities were not usurped by lesser men, whose intellectual capacities and moral integrity were deemed inferior. Yet many local gentlemen, obviously, did not agree with this assessment from the royal court about the social order; and if this had not been the case, the proclamation would not have been issued. Bacon's elite perspective here recalls his anxieties in 1589 concerning gentlemen who ignored their local civil duties in favour of 'voluntary communities', and also his anxieties about the political ramifications of knowledge unmediated by those of superior training and understanding. 'A Proclamation for Jurors' followed this hierarchical line of thinking so far as to propose that among the body of gentlemen from whom a sheriff should return a jury, 'the more principal persons' be chosen to hear 'the greatest causes': the greatest causes required the greatest men, since their integrity as jurymen was greater than that of the smaller gentry.[27]

The true organs of 'justicature', Bacon argued, included not only the king's Privy Council and his royal judges, but other royal officers as well: namely, the Attorney-General, the Solicitor-General, the Attorneys of the Courts of Wards and the Duchy of Lancaster, the lieutenants and deputy-lieutenants of the counties, the high sheriffs and the Justices of the Peace. And of the royal officers of 'justicature' in the counties, Bacon insisted

> that none be put into ... these commissions with an eye of favour unto their persons, to give them countenance or reputation in the

> place where they live, but for the King's service sake only; nor any to be put out for the displeasure of any great man: it hath been too often used, and it hath been no good service to his Majesty.[28]

Bacon wished to break the way justicature was organised in the counties. The Crown relied upon the unpaid services of landed gentlemen for the administration of justice, but this resulted in the unedifying spectacle of the Crown pandering to the (not so subtle) jostling for local prestige and power among landed gentlemen and lords: giving men 'countenance or reputation in the place where they live'. He hoped that men would be chosen solely on the basis of their loyalty to the king's service.

Bacon's announcement concerning rights of prior audience to judges' and serjeants' sons stemmed from his desire to establish royal judicial service as a privileged vocation. By demanding that gentlemen regard jury service as an honourable duty, by altering the manner in which the king's officers in the counties were chosen, and by especially privileging the sons of judges and serjeants, Bacon was seeking to discourage men from thinking primarily about local, personal ambitions and to foster the growth of a distinctive career: royal service. Bacon's own father and uncles had been among a very small circle of life-long Crown servants, and he had always believed royal service was his career. Wholly a royal government man, it was perfectly natural for him to believe that such ambitions were virtuous and such service was a noble vocation. Now he hoped to recruit others to this vocation by privileging the Crown's senior judicial servants and by protecting and insulating them from local interests and interference.

When Bacon wrote that the judges were best used 'for the government of the countries through which they pass', he was suggesting a major change in the role and the employment of the judges in the state.[29] Judges of the assize circuits had been accustomed to receiving reminders and exhortations upon specific topics which concerned the Privy Councillors (e.g., the failings in the administration of justice in certain districts, or the prevalence of vagabonds, and the need to set an example). But now, they were to be active in 'judicature' in the broadest sense of the word; they were to be fully integrated into the government of the counties on the king's behalf, as the 'best intelligencers to the king of the state of the whole kingdom'. In such a view of the judges' role in the labour of royal governance, there could be very little room for a bench of judges independent of their master. Bacon's assertion in 1612 that 'it is an Happie Thing in a *State*, when *Kings* and *States* doe often Consult with *Judges*; And againe, when *Judges* doe often Consult with the *King* and *State*', is not a merely platitudinous one.[30] In the light of his other claims about the judges, the full force of his most celebrated remarks can be better appreciated:

> Let no Man weakly conceive, that Just Laws, and True Policie, have an *Antipathie*: for they are like the Spirits, and Sinews, that One moves with the Other. Let *Judges* also remember, that *Salomons Throne*, was supported by Lions, on both sides; Let them be Lions, but yet Lions under the Throne; Being circumspect, that they doe not checke, or oppose any Points of *Soveraigntie*.[31]

Recall as well Bacon's speeches to the judges before they went on circuits, in which he called them the 'planets of the Kingdom' and likened their circuits to the 'rivers of Paradise, they go to water the whole kingdom'.[32] The planets could not defy the prime mover; the rivers of paradise could not rebel; and the 'lions' of justice could not emerge from beneath Salomon's throne. No less than the king himself, '*Judges* . . . imitate God, in whose Seate they sit'; they were not free to act as they saw fit.[33]

What Bacon had suggested about 'purging' the law reports would curtail the powers of the judges. He had pointed out that the 'substance of the Common Law' was the 'Judicial Acts' (found in the 'Year Books or Reports') and, as we have seen, common lawyers believed that the common law was an all-embracing body of knowledge, various truths from which had been captured (albeit veiled) in the reports, and were made clear only to those most learned in the law. Bacon proposed to remove all the ambiguities in one fell swoop. Even supposing those employed were learned in the law, and scrupulous, the project would take from the common lawyers all those cases and judgements which Bacon's committee thought 'obsolete and of no use', 'iterations' and 'seeds of incertainty': that is to say, just the sort of thing presently sorted out by judges.[34] If one shared the typical common lawyers' views about the law, and about their procedures and their necessary role in discovering the hidden law, then such a project would be grotesque and heretical. Yet Bacon, clearly, did *not* share their views. His plans for the law reports would have been little short of a revolution in the law. 'Thus much for the time past', indeed.

Bacon singled out Coke's *Reports* for special attention. In one respect, Bacon's attacks could be seen as personal ones: the two men had become bitter rivals in the 1590s, not only as ambitious rival lawyers (witness the contest over the Attorney-Generalship in 1594, and Bacon's failure to win the Solicitor's post – which he later, and improbably, blamed on Coke's machinations) but also as rival suitors in 1597–8 for the hand of Lady Hatton, a wealthy young widow, (Coke won). An incident in the Court of Exchequer in 1601, when Coke unleashed furious abuse upon Bacon, and the content of Bacon's subsequent letters to Robert Cecil and to Coke himself suggest just how deeply Coke resented Bacon, with his favoured but supernumerary status as 'Queen's Counsel Extraordinary'. The incident can only have reinforced the deep personal loathing with which they regarded each other.[35] Yet Bacon's attacks on Coke's *Reports* stem from greater considerations on his part. Coke's *Reports* had achieved favoured status within the profession immediately after he began publishing them in 1600. They were easily the most comprehensive of the modern reports, and their author was already highly respected within the profession. The effect of his *Reports* was to increase even further Coke's influence with the common lawyers and over the common law. Bacon's attack upon the integrity of these *Reports* was not just another instance of bitter rivalry between the two men. Given Bacon's audacious plans for law reform, we can see that there was much more at issue here. For Edward Coke was the principal voice of opposition towards King James' assertions of his royal prerogatives. Not only was his a loud voice, but a powerful one – especially from 1606, when he became Chief Justice of the Common Pleas.[36] Francis Bacon felt Coke was a dangerous obstacle in 'the King's causes', and he recommended to the king that he be muzzled. In a letter to King James in late summer 1613, Bacon argued for removing Coke from the Common Pleas on the grounds that 'how necessary it is for your Majesty to strengthen your service among the Judges by a Chief Justice which is sure to your prerogative, your Majesty knoweth'. He went on to argue,

> the removal of my Lord Coke to a place of less profit (though it be with his will) yet will be thought abroad a kind of discipline to him

> for opposing himself in the King's causes, the example of which whereof will contain others in more awe.[37]

The 'others' in question were the other judges. Thus Bacon's concern to reduce Coke's influence was not simply due to personal enmity, but because he saw Coke as corrupting the loyalty of the judges in 'the King's causes'. James believed his prerogative powers were huge but irritatingly undefined in the common law; he desired (at the least) that his courts support him in any legal tests of his claims.[38] Coke continued to be difficult, and his attack on the legal authority of the Chancery in 1616 swiftly brought his suspension from office.[39] It is striking that King James not only suspended Coke but insisted that he devote his enforced leisure to a revision of his *Reports*, and then submit them for scrutiny to Lord Chancellor Ellesmere and Bacon. In Coke's *Reports*, Bacon had identified no fewer than seventeen major 'innovations introduced into the Laws and Government', each of which undercut royal powers and the authority of courts other than the common law courts.[40] The Chief Justice, Bacon wrote to the king, 'being neither liberal, nor affable, nor magnificent, he hath made himself popular by design only, in pulling down government'.[41] Coming from the king's Attorney-General, this accusation was little short of a charge of sedition. While Bacon was pleading with the king for permission to 'prune' law reports, Coke (in Bacon's eyes) was trying to use them against the king's interests. Whenever he thought of Edward Coke, Bacon reconfirmed his belief that law reports were politically sensitive in themselves and never should be produced by 'voluntaries'.

Bacon wanted to compile from the ancient royal records a book he called *De antiquitatibus juris*. Searching, weighing, selecting and summing 'with judgement' would take a considerable length of time and, presumably, another hand-picked legal committee. Yet unlike the Year Books and the modern reports, the records of royal governance were not a public body of documents; they were stored haphazardly, and only a few lawyers and antiquaries sought permission to search the records. But lawyers doing this could often find persuasive ammunition for their cases amid the discarded records of medieval royal government, and notable amongst these lawyers was Edward Coke, whose actual or fictitious discoveries in the ancient records were a central part of his *Reports*. Bacon was proposing to 'set down and register' an official book of 'ancient records . . . acts of parliament', and the like – just as he had for the Year Books and reports – and he was proposing it for just the same reasons. Unlike Coke, who was always striving to enlarge the competence and jurisdiction of the common law, Bacon claimed that the common law did not automatically contain anything that one might unearth in the ancient royal records, and that the king's hands were not to be bound by pointing to the actions (as opposed to the statutes) of his predecessors. Just like law reporting in general, such antiquarian techniques were dangerous in the hands of 'voluntaries'. Dangerous as well was the view that underpinned such painstaking labour: namely, that one could discover hidden law by looking at the mouldering documents of earlier royal governance.

Bacon's desire to 'prune' the book of statutes and the books of law reports and to make a book of selections from the ancient records stemmed not from any simple-minded love for tidiness. What was at stake with regard to these 'digests' was control of the law, and thus the possibility of a further consolidation of royal power in the state: did the king rule, or did the judges? Bacon knew that the law reports were the key to any knowledge of the law. It was on the basis of the ancient, enigmatic Year Books and the various modern reports that lawyers could twist and turn for an 'issue of law', and the serjeants and judges discover hidden law from the most recalcitrant materials. If Bacon's king was not simply to preside over the state but to *rule* it, he would need some control over the process of arriving at judicial decisions about the common law, and much might be done towards this goal by carefully 'pruning' the law reports. Bacon could reasonably expect King James to give his plans a sympathetic ear; after all, some judges (and particularly Coke) were not only opposing the king in celebrated cases concerning his royal prerogatives, but they were claiming the law required their opposition – and doing so on the basis of ancient statutes, ambiguous documents, and dubiously constructed reports. It is highly likely that Bacon had Edward Coke in mind when in late 1612 he wrote:

> Judges ought to remember that their Office is *Ius dicere*, and not *Jus dare*; to interprete law and not to make law, or give Law; Else it will be like the presumption of the Church of *Rome*, which under pretext of exposition of Scripture, usurpeth and practiseth an authority to

> adde and alter; and to pronounce that which they doe not finde, and by colour of Antiquity to introduce novelty.[42]

Bacon did not hide his anxiety about the judges' pronouncements: 'Particular and positive learnings of laws, do easily decline from a good temper of justice, if they be not rectified and governed by such rules.'[43] The construction of a text *De regulis juris* would help to control the judges, by providing a catalogue of *legum leges* as a public standard with which to measure individual judgements. The effect would be to reduce the amount of personal experience that a judge would need to call upon, and to devalue the nearly mystical 'learnedness-in-the-Law' which judges like Coke used to add authority to their pronouncements. If *legum leges* could be discovered and published, then the law would indeed be more certain and much help be given to 'soundness of judgement', and, what is more, the judges themselves would lose much freedom to make idiosyncratic judgements (notwithstanding Bacon's willingness to deliver this knowledge in 'disjoined aphorisms which doth leave the wit of man more free to turn and toss'). If his proposals to muzzle the judges had been adopted, it is certain that the effect would have been precisely what Bacon claimed: 'the amendment . . . of the very nature and complexion of the whole law'. Even after his fall from office in 1621, Bacon planned to return to this project if he was restored:

> For my Pen. If Active, 1. The recompiling of laws. 2. This disposing of wards and generally education of youth. 3. Limiting the jurisdiction of courts, and prescribing rules for every one of them.[44]

It is hardly surprising that much of Bacon's programme for legal reform was voiced in private letters which were made public only after his death.

Bacon's purpose in planning three 'digests' (a new statute book, official law reports, and a book *De antiquitatibus juris*) was to emasculate the common law as practised by his contemporaries and to subordinate the profession to its royal master. His 'digests' were to be generated by committees hand-picked by the king (or by a nominee such as Bacon), and once lawyers and novices began to use them rather than the old Year Books, for instance, then there would be a swift change in the professional memory of the common lawyers. To be called 'learned in the law' hereafter would mean something very different than before. It would mean the study and mastery not of the Year Books but of the official reports (from which all 'seminaries of doubts and incertainties' had been removed) and the contemplation of an official body of *regulae juris*. Where then would be the 'common erudition' of the English common lawyers?

The *scientia* of the common law (i.e., the disciplined study of this body of knowledge) which underpinned Bacon's *Maxims of the Law* was not that of contemporary common lawyers. What Bacon was hoping to achieve by creating and repeatedly supplementing a body of *regulae juris* was a code, and he was perfectly explicit that it would be modelled on the law code of the Romans. It is worth recalling Bacon's speech commending just such compilations to the parliament of 1593.[45] Bacon was engaged in exactly this enterprise: he aimed to appoint committees, to 'recall' the old laws, and to 'set forth' new ones. He spoke repeatedly of the Romans and their laws, and believed their law was much better than the existing civil law:

> The Emporor Justinian, beinge rightly called *ultimus Imperatorum Graecorum* ... chose it for a monment and honor of his goverment, to revisite the Roman Lawes, and to reduce them from the infinit volumes and muche repugnancye, and incertaynety into on competent and uniforme corps of lawe. Of which matter himselfe doth speake gloriously, and yet aptly callinge it *Proprium et sanctissimum templum Justiciae consecratum*. A worke of greate excellency indede as may well appeare in that France Italie and Spayne, which longe agoe have shaken of the yoake of the Romayne Empyr, doe yet never the lesse continue to use the Pollicye of that lawe, but more excellent had the worke bin save the more ignorant and obscure tyme undertoke to correct the more learned and flourishinge tyme.[46]

It was useful for Bacon to claim that the so-called 'Roman' or 'civil' law of Justinian's day (as well as that of his own time) was a degenerate form of the Romans' own law. It helped him argue that he was not proposing to introduce a form of civil law – a disavowal which he felt obliged to make on many occasions, but one which was surely disingenuous.[47]

The principal areas of concern among Englishmen who feared the civil law were (on the one hand) trial procedure and court process in general and (on the other) the powers of the ruler.[48] They feared that common law courts and trial procedures would be swept away, and that the king would be given an untrammelled authority in the state.[49] A clear example of this widespread unease can be found in the reaction of the parliament of 1607 to a book by a doctor of the civil laws, in which the powers of the king were described in terms parliament thought dangerous.[50] In such a climate, Bacon could hardly have said he wished to introduce the civil law into the common law courts, and it is not surprising that he wrote no documents explicitly stating that he wished to change common law process and trial procedures. Nevertheless, we have seen just how suspicious he was of the common law and why he feared its sages. Yet once Bacon had implemented his reforms, the *science* of the common law would be indistinguishable in practice from the science of the civil lawyers. *De significationibus verborum* and *De regulis juris*, after all, were the crucial final titles in the Roman *Digest*.[51] Having established official legal texts, Bacon would not need to do anything more than wait. The old common law Year Books would be valid no longer. Over a period of years, common law trial procedures and court process would be adjusted to the new rules and the 'pruned' reports: anything done in contradiction to them would not be 'common' law.[52]

Bacon's various legal reform proposals, then, were tactical elements in three mutually supporting campaigns: the promotion of a distinctively royal elite for judicature; the curtailing of the independence of the judges; and the imposition of a new science for English law. We are now in a position to examine the broad picture, and to consider just what it was that Bacon expected to achieve by means of his legal reforms. In other words, these audacious campaigns were part of a greater programme.

Prerogative and empire

Bacon's sweeping legal reforms had profound implications for the structure of governance in the state. His plans to codify the rules and bases of judicial decisions and to alter the role of the judges in royal governance would have had the effect of severely reducing the independence of the judges. Yet this effect could only be relative to the independence possessed by some other person or persons who held rights to make law in the realm.[53] Given Bacon's desire to curtail the powers of the judges ('their Office is ... not to *Make Law*, or *Give Law*') and given as well his scorn for parliament's pretensions to authority ('more properly a Council ... than a Court'), then the obvious legislative power in a monarchical state was the king alone.

Bacon repeatedly spoke in glowing terms about the powers of the Crown, the 'flowers of the Crown', the 'royal prerogatives'. And on most of the occasions when he did so, he was engaged in strenuously attacking suggestions that there was legitimate ground for grievance against the royal exercise of the prerogative.

> For the prerogative royal of the Prince, for my own part I ever allowed of it; and it is such as I hope I shall never see discussed. The Queen, as she is our Sovereign, hath both an enlarging and restraining liberty of her Prerogative: that is, she hath power by her Patents to set at liberty things restrained by statute law or otherwise: secondly, by her Prerogative she may restrain things that are at liberty.[54]

In 1606, in the course of a protracted suit before the judges concerning the rightful jurisdiction of the Council for the Welsh Marches, Bacon prepared a paper denying any intermediate source of authority in the state:

> The King holdeth not his prerogatives of this kind mediately from the law, but immediately from God, as he holdeth his Crown: and though other prerogatives by which he claimeth any matter of revenue, or other right pleadable in his ordinary courts of Justice, may be there disputed, yet his sovereign power, which no Judge can censure, is not of that nature; and therefore whatsoever partaketh or dependeth thereon, being matter of government and not of law, must be left to his managing by his Council of State.

During the same suit, Bacon attacked all assertions by lesser justices of any authority independent of the king. In doing so, he reiterated his central motives for such arguments: first, his fears of a dissolution of 'proper' social and political order, and secondly, his demand that the state be much more than a monarchy in name only:

> Monarchies in name do often degenerate into Aristocracies or rather Oligarchies in nature by two insensible degrees. The first is when prerogatives are made envious or subject to the constructions of laws; the second when law as an oracle is afixed to place. For by the

> one the King is made accomptable and brought under the law; and by the other the law is over-ruled and inspired by the Judge; and by both all tenures of favour, privy Council, nobility and personal dependances (the mysteries that keep up states in the person of the Prince) are quite abolished, and magistracy enabled to stand of itself.

He argued further that the common law and the machinery of judicature were the inventions of English kings:

> As Bracton saith *Rex habet ordinariam potestatem et omnia jura in manu sua*. And their practice in all times doth confirm as much. For King Alfred first divided the land into shires and provincial law days. William the Conqueror brought in the Exchequer, and kept the Chancery and Common Pleas at his Court. Henry the 3rd erected the Admiralty and the Duchy. Edward the 4th the Councils of Star chamber and the Marches. Henry the 8th set up the Courts of Requests, of Wards and Liveries, and the Council at York. And though some of these were since approved by Statute Law, yet the author and life giver of them all was the prerogative of our Kings, which must of necessity be impeached in all, if denied in any.[55]

In 1608, in the great case of the *Post-nati*, Bacon argued before all the judges on the central issue of law 'whether a child, born in Scotland since his Majesty's happy coming to the crown of England, be naturalised or no?' Much of his argument turned on the powers of the monarch and the nature of kingship in England. At one point he declared:

> As it was said by a principal judge here present when he served in another place ... that he would never allow that queen Elizabeth ... should be a statute Queen, but a common-law Queen: so surely I shall hardly consent that the King shall be esteemed or called only our rightful sovereign, or our lawful sovereign, but *our natural leige sovereign*; as acts of parliament speak: for as the common law is more worthy than the statute law: so the law of nature is more worthy than them both.[56]

Bacon was powerfully convinced that the achievement of an untrammelled authority in the state by the monarch was 'natural' and supremely valuable for civil society. He worked consistently towards this end. Because he believed such an achievement could easily degenerate into 'aristocracies' or 'oligarchies', he felt the need for constant vigilance and counter-attacks against what he saw as attempts to undermine royal control of the machinery of governance. These worries, typically voiced in his assertions of the king's prerogatives, lay behind Bacon's blistering assault in 1612 on the stubborn opinions of the lawyer James Whitelock:

> Whitelocke had affirmed and maintained ... that the King cannot, neither by commission nor in his own person, meddle with the body, goods, or lands of his subjects, but only by indictment, arraignment, and trial, or by legal proceedings in his ordinary Courts of Justice, laying for his ground the statute of *Magna Charta, Nullus liber homo capiatur*, etc. which position in that general and indefinite manner was ... not only grossly erroneous and contrary to the rules of law, but dangerous and tending to the dissolving of government. First, for that *Lex Terrae* mentioned in the said Statute, is not to be understood only of the proceedings in the ordinary Courts of Justice, but that his Majesty's Prerogative and his absolute power incident to his sovereignty is also *lex terrae*, and is invested and exercised by the law of the land, and is a part thereof.[57]

After his personal humiliation and loss of office at the hands of the parliament of 1620–1, Bacon was equally (if not more) emphatic about the need for an unbridled exercise of royal authority in the state:

> Seeing therefore that laws necessarily change, it follows necessarily that in any commonwealth there is a certain power above the laws that can abolish them, and make them anew; and this power is clearly the most just and inward cause of the power of the laws in various empires.[58]

Not only must there be a 'power above the laws', but this power was a condition necessary for empire. Bacon was greatly excited by the accession of the king of Scotland to the English throne in 1603: 'Therefore, most high and mighty King, my most dear and dread sovereign lord ... now the corner-stone is laid of the mightiest monarchy in Europe.'[59] In drafting a royal proclamation shortly after James' accession, Bacon put into James' mouth his own justifications for a conquest of Ireland: 'to reduce that nation from their rebellion and revolt ... to populate, plant, and make civil all the provinces in that kingdom ... as we are persuaded it is one of the chief causes for the which God hath brought us to the imperial crown of these kingdoms'.[60] Later in 1603, speaking of the union of England and Scotland, Bacon suggested to King James that

> there hath been put in practice in government these two several kinds of policy in uniting and conjoining states and kingdoms; the one to retain the ancient forms still severed, and only conjoined in sovereignty; the other to superinduce a new form agreeable and convenient to the entire estate. The former of these hath been more usual, and is more easy; but the latter is more happy. For if a man do attentively revolve histories of all nations, and judge truly thereupon, he will make this conclusion, that there were never any states that were good *commixtures* but the Roman. Which because it was the best state of the world, and is the best example in the point, we will chiefly insist thereupon.[61]

Bacon argued in the parliament of 1606–7 that 'plantations' should be considered in terms of the 'greatness of Britain, and not just in terms of present 'utility':

> For greatness (Mr. Speaker), I think a man may speak it soberly and without bravery, that this kingdom of England, having Scotland united, Ireland reduced, the sea provinces of the Low Countries contracted [to us], and shipping maintained, is one of the greatest monarchies, in forces truly esteemed, that hath been in the world. For certainly the kingdoms here on earth have a resemblance with the kingdom of Heaven, which our Saviour compareth not to any great kernel or nut, but to a very small grain, yet such as is apt to grow and spread. And such do I take to be the constitution of this kingdom, if indeed we shall refer our counsels to greatness and power, and not quench them too much with considerations of utility and wealth.[62]

In a speech in 1619 which recalls the belligerence of his 1584 'Letter of Advice to the Queen' (if not its 'Godly' fervour), Bacon urged the king to begin the conquest of Spain's overseas colonies:

> Spain, his Majesty there, though accounted the greatest monarch of Christendom, yet if his estate be enquired through, his root will be found a great deal too narrow for his tops ... but for the Indies he were the poorest King of Europe ... Now of the fitness and honour of it to be an undertaking of [our] Majesty. He is the greatest islander of Christendom; therefore a navy is most proper to him. He is the Defender of the Faith, as well in understanding, learning, and godliness, as in title. Therefore the planting of the true Church there

> is a sacred work that even by office as it were belongs to him. He has a great and liberal mind: the Indies will afford him the means to exercise it.[63]

Bacon's commitment to empire was both public and utterly clear; for example, in his essay 'Of the Greatness of Kingdomes', he declared:

> Certainly in the great frame of Kingdomes and Commonwealths, it is in the power of Princes or Estates by ordinances and constitutions, and maners which they may introduce, to sowe greatness to their posteritie and succession. But these things are commonly left to chance.[64]

Yet Bacon was not prepared to leave the 'sowing of greatness' to chance. His legal reforms would provide the monarch with an unhampered sway over governance. This supremacy was worth struggling for not solely because it was 'natural' and valuable in itself, but because Bacon was convinced it was the necessary precondition of 'empire'. Bacon desired to establish not only an unquestioned monarchical power but an imperial Britain: *this* was Bacon's passion and his great project. His legal reforms and his reformed natural philosophy (schemes grand – but inexplicable – in themselves) were but means towards this political goal.

Natural philosophy and empire

Of greatest interest to us is Bacon's claim that natural philosophy, too, would support an imperial Britain. In 1592, he had declared 'the sovereignty of man lieth hid in knowledge' and, in 1595, he had argued for the 'conquest of the works of nature'. Even if these and many similar assertions were to be discounted as 'metaphor' and 'ornament', the frontispiece of *Instauratio magna* (1620) is an unequivocal declaration that natural philosophy would contribute to empire.

The frontispiece is dominated by a ship sailing towards a narrow strait between two great pillars. These are the mythical pillars of Hercules, and the ship is unmistakably a warship, and arguably of English design. At least from the time of Pindar in the sixth century B.C., the pillars of Hercules had been associated with the metaphorical limits of man's ambitions and explorations.[65] To emphasise the fatal repercussions of attempting to overreach a mortal man's limitations, the waters beyond the pillars were thought to teem with monsters, and in the Middle Ages the phrase *ne plus ultra* was attached to the visual emblem of the pillars. Yet a more immediate and imperial message can be found in Bacon's frontispiece. With the conquest of a vast overseas empire, the kings of Spain had adopted the pillars as their personal emblem, proudly changing the motto to *plus ultra*. What Bacon did was to appropriate the Spanish imperial emblem and to impose upon it an English ship of war. His frontispiece is a bold argument for an imperial policy. The text is devoted to Bacon's reforms of natural philosophy, rather than to proposals for territorial conquests, but the first message to the reader is unmistakable: a reformed natural philosophy will help bring an empire.

For the benefit of those who might miss the force of his arguments (and that of the frontispiece of *Instauratio magna*), Bacon decided to spell out his vision of an imperial monarchy sustained by natural philosophy; this is what *New Atlantis* is, in fact, about.[66]

New Atlantis has been often discussed as utopian literature and Bacon, certainly, was writing a fable.[67] Nevertheless, this text must also be considered in terms of centralised governance and a reformed, imperial, civil polity. The story is, in the first instance, about an imperial state, and only then about an ideal natural philosophical community. The political landscape of *New Atlantis* is often overlooked. The island state of Bensalem is a monarchy, ruled by an ancient line of Christian kings. Some 1900 years earlier, the great King Salamona, a 'divine instrument', the 'lawgiver of our nation', decreed that his many ships would be best employed not only by fishing and trading, but also 'by sailing into some small islands' and putting them 'under the crown and laws of this state'. King Salamona had a policy of overseas colonisation, and Bensalem was an island empire. As for the structure of governance of Bensalem, there is little said in the (incomplete) text preserved for us by Bacon's chaplain, but Bacon apparently planned to write a very great deal about the 'frame of Laws, or of the best state or mould of a common wealth'.[68] We do have a few important clues. The towns are ruled by governors, who take commands from the Crown itself. All the officials are state officers, and have state salaries, which made them independent of local influence and persuasions. Each of these men are easily distinguished from each other (and from the people) by their robes and badges of office. Here are striking similarities to Bacon's proposals for the reform of England's governance and the colonisation of the West Indies. Striking, too, is the fact that the first great king of Bensalem bears the very name, 'Solomon', which Bacon repeatedly applied to King James.

Given that Bacon had planned to describe the structures of governance in Bensalem, it is significant that the first major state institution discussed in *New Atlantis* was 'Solomon's House'. Solomon's House was 'dedicated to the study of the Works and Creatures of God':

> The end of our Foundation is the knowledge of Causes, and secret motions of things, and the enlarging of the bounds of Human Empire, to the effecting of all things possible.

The foundation had two purposes: *scientia* and the enlarging of 'Human Empire'; the effect would be to bring 'the effecting of all things possible' within the grasp of the Brethren of Solomon's House. This 'noblest foundation that ever was upon the earth' was not secluded from the world: it was 'the lanthorn of the kingdom'. Although they were called 'Brethren', these men were not to be confused with monks, who disavowed the world; on the contrary, Bacon made a point of describing their magnificent robes and bearing: they were great noblemen, an elite class of state pensioners.

At the disposal of the brethren are a vast complex of rooms, houses, caves, ponds, gardens, parks, brewhouses, furnaces, dispensaries and the like for their discoveries into 'the knowledge of Causes and secret motions of things'. Yet these 'preparations and instruments' are utterly conventional. Sheer magnitude apart, this great assemblage of research sites is nothing other than what Bacon had suggested thirty years earlier as the 'Councillor Advising the Study of [natural] Philosophy', namely, the construction of a full range of those 'works and monuments' (cabinets, museums, parks and 'engine-rooms') built by many Renaissance princes. The novelty of Solomon's House lies not in these 'instruments', but in the very structure of the *intellectual* labour conducted by the brethren. This labour is rigidly organised in accordance with Bacon's conviction that 'true knowledge and certain proofs' demand the mediation of experts. Solomon's House displays a hierarchical institutional structure for the eventual production of knowledge about the nature of things.

New knowledge about the natural world is produced through a chain of committees, each of which process information provided by an earlier group of brethren. At the beginning of this chain are twenty-one of the thirty-six brethren, who are devoted to collecting information (largely concerning 'experiments') and who principally collect it from *books*:

> We have twelve that sail into foreign countries ... who bring us the books, and abstracts, and patterns of experiments of all other parts. These we call Merchants of Light.
>
> We have three that collect the experiments which are in all books. These we call Depredators.
>
> We have three that collect the experiments of all mechanical arts; and also of all liberal sciences; and also of practices which are not brought into arts. These we call Mystery-men.
>
> We have three that try new experiments, such as themselves think good. These we call Pioneers or Miners.[69]

The information gathered by these groups was then given to a committee of three 'Compilers', who 'draw the experiments ... into titles and tables, to give better light for the drawing of observations and axioms out of them'. After 'divers meetings and consults' of all the brethren concerning the 'former labours and collections', three 'Lamps' decide upon novel experiments to be performed by three 'Inoculators', who report their results to three 'Interpreters of Nature' . These men 'raise the former discoveries by experiments into greater observations, axioms, and aphorisms'. Then, three 'Dowry-Men, or Benefactors'

> looking into the experiments of their fellows ... cast about how to draw out of them things of use and practice for man's life, and knowledge as well for works as for plain demonstrations of causes, means of natural divinations, and the easy and clear discovery of the virtues and parts of bodies.

This organisation can be displayed in a chart:

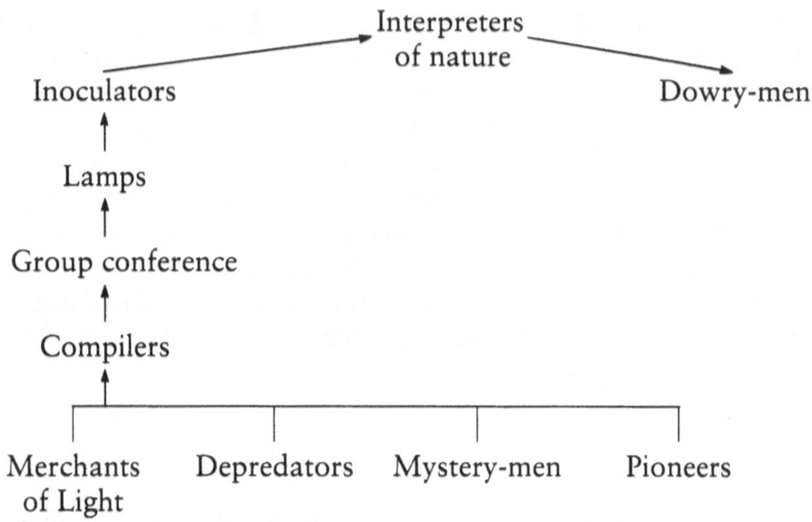

After the 'Dowry-Men' have done their work, there are more group 'consultations' in order to decide

> which of the inventions and experiences which we have discovered shall be published, and which not: and all take an oath of secrecy, for the concealing of those which we think fit to keep secret: though some of those we do reveal sometimes to the state, and some not.

The fruits of the labours of the brethren are not automatically made public, and the issue of publication is important enough to require a full council. Why some knowledge is reserved is never discussed, but we may assume it is because Bacon believed knowledge can be socially and politically dangerous. Not surprisingly, given Solomon's House is a state institution and the brethren are state officials, the state alone is offered some of this forbidden fruit. The knowledge which they do choose to make public, however, they deliver only when on 'circuits, or visits of the divers principal cities of the kingdom':

> We do publish such new profitable inventions as we think good. And we do also declare natural divinations of diseases, plagues, swarms of hurtful creatures, scarcity, tempests, earthquakes, great inundations, comets, temperature of the year, and divers other things: and we give counsel thereupon what the people shall do for the prevention and remedy of them.

They only publish 'such new profitable inventions as they think good', yet the criteria for 'goodness' we can only guess about. What *is* clear is that the people remain passively in a state of ignorance about natural knowledge. The brethren choose for them certain 'inventions', the end-products of their intellectual labours, rather than give out any fertile insights into the 'causes and secret motions of things'. Even after centuries, the brethren have not removed from the people diseases, plagues, scarcity and so on, and they have not enlightened the people (who, presumably, are fearful) about the nature and causes of comets. This state of affairs may well be deliberate, for it guarantees not only that the brethren have a paternal civil role ('giving counsel what the people should do for . . . prevention and remedy') but also that they are the only people in Bensalem who possess the *scientia* of natural knowledge. Furthermore, they do not dispense their counsel and inventions very often. The last time one of the brethren visited the city where our narrator is staying was twelve years earlier. This is indeed a highly restrictive policy about the dissemination of natural knowledge.

Not only his few remarks about the civil arrangements of Bensalem and the description of Solomon's House within this state reflect Bacon's concerns about governance and a reformed civil polity. So, too, does the very choice of a title for this work, an unmistakable reference to Plato's story of Atlantis. Plato discussed Atlantis in the *Critias* and (to a lesser extent) in the *Timaeus*, both of which belong to a sequence of dialogues about politics beginning with the *Republic* and ending with the *Laws*. Atlantis functioned in these dialogues as a rival to Athens, and as a foil to the discussions of an ideal state. Bacon chose to entitle his description of the kingdom of Bensalem *New Atlantis* not only so his readers would make an advantageous comparison between Bensalem's polity and destiny with that of Plato's (flawed and doomed) kingdom of Atlantis, but so as to emphasise the parallel between the function of the two stories in the larger concern of both Plato and Bacon with the 'frame of Laws, or the best state or mould of a commonwealth' (as Bacon's chaplain put it).

Other parallels reinforce this reading. According to Plato, an Egyptian priest had told the Greeks that the great navy of ancient Atlantis, intent upon conquering all the other nations, was defeated by an Athenian fleet:

> And then it was ... that the manhood of your state showed itself conspicuous for valour and might in the sight of all the world. For it stood pre-eminent above all in gallantry and all warlike arts, and acting partly as leader of the Greeks, and partly standing alone by itself when deserted by all others, after encountering the deadliest perils, it defeated the invaders and reared a trophy; whereby it saved from slavery such as were not yet enslaved, and all the rest of us who dwell within the bounds of Heracles it ungrudgingly set free.[70]

The parallel is with the resistance of Protestants (led and funded by England) against Spain and the Catholic League, with the English defeat of the Spanish Armada, and with Bacon's urging 'the planting of the true Church' in Spain's colonies. Plato, furthermore, had declared that the story of the polity and the defeat of Atlantis was brought to his contemporary Greeks by none other than Solon, the great lawgiver of Athens (whose reforms led to Athens' imperial greatness). Surely a parallel lies between Solon and Bacon, both lawgivers, and both the bringers of stories justifying an imperial destiny for their countries.

In outlining the imperial state which Bacon desired to institute in Britain, we have drawn upon two unlikely topics, his plans for legal reform and his plans for an institute for investigating the secrets of nature. Both projects have shown the same political concerns and convictions at work. In truth, his legal reforms and his reformed natural philosophy only make sense within the context of these prior political concerns and convictions. We are now in a position to explore the details of the parallels between Bacon's reformed law and a reformed natural philosophy.

6

A reformed natural philosophy

> The laws of heaven and earth . .
> are the subject of philosophy.
>
> *Advancement of Learning* (1605), *Works*, III, p. 379

The administration of the world

Bacon's perspective was always that of a statesman; he believed problems of knowledge were properly part of a statesman's concerns; and he believed a reformed natural philosophy would be a splendid support for the imperial state he desired to see established. Such a natural philosophy would (he said) bring forth knowledge of 'the causes and secret motions of things', and produce inventions and useful knowledge for the state and the people.[1] Nevertheless, Bacon argued repeatedly that inventions and the advantages they would bring to the commonweal were *never* to be mistaken for 'the true ends of knowledge'.

> It is not the pleasure of curiosity, nor the quiet of resolution, nor the raising of the spirit, nor victory of wit, nor faculty of speech, nor lucre of profession, nor ambition of honour or fame, nor inablement for business, that are the true ends of knowledge; some of these being more worthy than other, though all inferior and degenerate: but it is a restitution and reinvesting (in great part) of man to the sovereignty and power (for whensoever he shall be able to call the creatures by their true names he shall again command them) which he had in his first state of creation. And to speak plainly and clearly, it is a discovery of all operations and possibilities of operations from immortality (if it were possible) to the meanest mechanical practice.[2]

Bacon made many such statements, and although they often included an allusion to repairing the damage of the Fall from Paradise, it should not be assumed that Bacon was seriously motivated by millenarian enthusiasm in making them. These are justifications and defences of his proposals for natural philosophy;

quite apart from common piety, Bacon clearly felt the need to guard his flank against criticism by those who would condemn him and his programme for natural philosophy as diversionary or impious. Equally typical among his bolstering arguments was the following caution:

> A little natural philosophy and the first entrance into it inclines men's opinions to Atheism; but on the other hand much natural philosophy and a deeper progress into it brings men's minds about again to religion.[3]

For Bacon, 'atheists' were not those who believed there was no God, but the adherents of what he viewed as a specific philosophical sect ('the Schoole of Leucippus, and Democritus, and Epicurus')[4] who denied there was a God over and above the creation, and who identified Him with fundamental principles *within* the universe. Bacon's major objection to this lack of 'religion' was that it denied God the 'government' and 'administration' over His universe:

> Epicurus ... affirmed there were blessed natures, but such as enjoyed themselves, without having respect to the government of the world ... And although he had the confidence to denie the administration; he had not the power to deny the nature.[5]

Bacon's God was an absolute monarch, the divine lawgiver and judge, whose laws governed the world. The crowning importance of knowledge of the natural world was, for Bacon, its ability to teach us these laws, 'the true rules of policy'. He made this abundantly clear in 1603 in a highly important letter to the new king:

> There is a great affinity and consent between the rules of nature, and the true rules of policy: the one being nothing else but an order in the government of the world, and the other an order in the government of an estate ... I conclude there is ... a congruity between the principles of Nature and Policy.[6]

Bacon went on to say that a 'resemblance' existed between the kingdoms on earth and that of heaven, and to argue that the best natural philosopher would be the best royal councillor, because by his labours he would acquire the knowledge by which best to run a civil society:

> I do not find it strange (excellent King) that when Heraclitus, he that was surnamed the obscure, had set forth a certain book which is not now extant, many men took it for a discourse of nature, and many others took it for a treatise of policy and matter of estate ... And therefore the education and erudition of the kings of Persia was in a science which was termed by a name then of great reverence, but now degenerate and taken in ill part: for the Persian 'magic', which was the secret literature of their kings, was an observation of the contemplations of nature and an application thereof to a sense politic; taking the fundamental laws of nature, with the branches and passages of them, as an original and first model, whence to take and describe a copy and imitation for government.

Seeing himself as one of the 'aforesaid instructors', Bacon made an avowal to his king:

> This knowledge, then, of making the government of the world a mirror for the government of a state, being a wisdom almost lost (whereof the reason I take to be because of the difficulty for one man to embrace both philosophies) I have thought good to make some proof ... to revive ... For truly (as has been said) it is a form of discourse anciently used towards kings; and to what king should it be more proper than to a king that is studious to conjoin contemplative virtue and active virtue together?

In the light of these declarations, we can better appreciate Bacon's remarks as the 'Councillor advising the Study of Philosophy' in his Gray's Inn masque of 1594:

> Antiquity, that presenteth unto us in dark visions the wisdom of former times, informeth us that the governments of kingdoms have always had an affinity with the secrets and mysteries of learning. Amongst the Persians, the kings were attended on by the Magi. The Gymnosophists had all the government under the princes of Asia; and generally those kingdoms were accounted most happy, that had rulers most addicted to philosophy. The Ptolemies in Egypt may be for instance; and Salomon was a man so seen in the universality of nature that he wrote a herbal of all that was green upon the earth.[7]

Bacon reaffirmed these conclusions in *Of the Proficience and Advancement of Learning, Divine and Humane*, which he published in 1605. 'Both in Egypt, Assyria, Persia, Grecia, and Rome', he argued, 'the same times that are most renowned for arms, are likewise most admired for learning, so that the greatest authors and philosophers and the greatest captains and governors have lived in the same ages.'[8] The 'great affinity' between the rules of nature and those of civil government was similarly emphasised here:

> Is not the ground, which Machiavel wisely and largely discourseth concerning governments, that the way to establish and preserve them, is to reduce them *ad principia*, a rule in religion and nature, as well as in civil administration? Was not the Persian Magic a reduction or correspondence of the principles and architectures of nature to the rules and policy of governments? ... Neither are these only similitudes, as men of narrow observation may conceive them to be, but the same footsteps of nature, treading or printing upon several subjects or matters.[9]

We now can understand several of Bacon's favourite arguments and slogans for his new natural philosophy. One Scriptural passage in particular (Proverbs 25:2), which he attributed to 'Salomon the king', stands out in Bacon's writings about natural philosophy:

> 'The glory of God is to conceal a thing, but the glory of the king is to find it out'; as if, according to the innocent play of children, the Divine Majesty took delight to hide his works, to the end to have them found out; and as if kings could not obtain a greater honour than to be God's playfellows in that game; considering the great commandment of wits and means, whereby nothing needeth to be hidden from them.[10]

Similarly, Bacon viewed the Fall ('the ambitious and proud desire of moral knowledge of good and evil') as a revolt from God, by which man gave 'laws unto himself', that is to say, imperfect and misguided civil governance.[11] Bacon's ambition to recover the ancient learning about the 'great affinities' lead him to declare at the beginning of *Instauratio magna* (1620):

> Being convinced that the human mind makes its own difficulties ... he thought all trial should be made, whether that commerce between the mind of man and the nature of things ... might by any means be restored to its perfect and original condition.[12]

This political ambition underlay his declaration in *Novum Organum* (1620): 'I am building in the human understanding a true pattern of the world [*exemplar mundi*], such as it is in fact, not such as a man's reason would have it to be.'[13]

Given his convictions about the affinity between the laws of nature and those of the civil polity, and about the primacy of discovering the latter, it was completely appropriate that Bacon should entitle the two books of *Novum Organum* as *Aphorisms concerning the Interpretation of Nature and the Kingdom of Man.* A proper study of natural philosophy, he believed, was not simply the means of recovering 'that commerce between the mind of man and the nature of things', but it was man's key to the knowledge of truly potent civil governance. This is why Bacon would declare repeatedly, 'knowledge itself is power', and why we can gloss his dictum as 'natural knowledge can magnify the powers of the state'.[14]

The interpretation of nature

Uncovering God's laws for civil society required the preliminary labour of recovering 'that commerce between the mind of man and the nature of things', a programme first represented by a memorandum of 1603. Bacon was convinced that it required the coordination of many helpers, the eradication of their intellectual presuppositions, and a jealous control over the release of information about the natural world:

> Now for my plan of publication: those parts of the work which have it for their object to find out and bring into correspondence such minds as are prepared and disposed for the argument, and to purge the floors of men's understandings, I wish to be published to the world and circulate from mouth to mouth: the rest I would have passed from hand to hand, with selection and judgement. Not but I know it is an old trick of impostors to keep a few of their follies back from the public which are indeed no better than those they put forward: but in this case it is no imposture at all, but a sober foresight, which tells me that the formula itself of interpretation, and the discoveries made by the same, will thrive better if committed to the charge of some fit and selected minds, and kept private.[15]

Just as in Solomon's House, 'some fit and selected minds' alone were permitted to discover natural knowledge. Given Bacon's long-standing suspicions about 'voluntaries', his demand that all the investigators be devoted servants of the state is perfectly understandable. Because 'knowledge is power', it was potentially politically dangerous, since it provided the key not only to material prosperity, but (as they were God's) to the true and irresistible hidden principles of civil governance as well.

The need to 'find out... such minds as are prepared and disposed for the argument' and 'to purge the floors of men's understandings' was a central concern in a considerable number of Bacon's writings, both published and abandoned. These include 'In Praise of Knowledge' (1592), 'Valerius Terminus: The Interpretation of Nature' (*c.* 1602), 'De interpretatione naturae prooemium' (1603), *Advancement of Learning* (1605), 'Temporis partis masculus' and 'Redargutio philosophiarum' (1608), 'Descriptio globis intellectus' (1612) and *De dignitate et augmentis scientiarum* (1623). Such texts were exhortations (some solemn, others irascible in tone) designed to bring men 'into correspondence' with Bacon, principally by criticising the existing state of learning and the assumptions upon which it rested. Bacon never attracted the support he desired, however, and he continued to work alone. The survival of a large number of manuscripts concerning 'the formula of interpretation' and other researches bear witness to solitary inquiry conducted over many years.[16]

It was not until 1620, when sixty years of age, that Bacon openly revealed the full scope of his programme in *Instauratio magna*. The work, he declared, 'is in six parts':

1. The Divisions of the Sciences.
2. A New Instrument, or Disclosures about the Expounding of Nature.
3. The Phenomena of the Universe, or Natural and Experimental History for constructing Philosophy.
4. The Ladder of the Intellect.
5. Forerunners, or Anticipations of the New Philosophy.
6. The New Philosophy, or Practical Knowledge.[17]

The first part was to provide 'a summary or general description of the wisdom or doctrines which the human race at present possesses'. This description was a highly critical one: 'I take into account not only things already invented and known, but likewise things omitted which ought to be there ... For I do not propose merely to survey these regions in my mind, like an augur taking auspices, but to enter them like a general who means to take possession.'[18] Clearly this labour is the preparing and disposing of men's minds of which he had written in 1603.

As for the second part, a 'New Instrument', Bacon declared:

> Having thus coasted beyond the ancient arts, the next point is to equip the intellect for passing beyond. To the second part therefore belongs the better and more perfect use of human reason in the inquisition of Things, and the true helps of the intellect: that thereby (as far as the condition of mortality and humanity allows) the intellect may be raised and exalted, and made capable of overcoming the difficulties and obscurities of nature. The art (which I call *Interpretation of Nature*) which I introduce with this view is a kind of logic; though the difference between it and the ordinary logic is great, indeed immense.

This new art included two major elements. One was the 'expurgation of the intellect to qualify it for dealing with truth', and the other was the display of a manner of demonstration for natural philosophy superior to traditional logic. This, too, conforms with Bacon's remarks of 1603, as this 'second part' was designed 'to purge the floors of men's understandings'.

> But I design not only to indicate and mark out the ways, but also to enter them. Therefore the third part of the work embraces the Phenomena of the Universe; that is to say, experience of every kind, and such a natural history as can be a foundation for constructing philosophy. For in truth an excellent way of demonstration, or form of interpretation, may defend and preserve the mind from errors and lapses, but it cannot furnish the material of knowledge itself.

Bacon believed that natural and experimental histories would be crucially important for the success of his new natural philosophy. These would be of a new kind 'in end or office, in bulk and composition, in exactness, in selection and in ordering'.

The fourth and fifth parts of the work, examples (on the one hand) of Bacon's manner of inquiry and (on the other) of his tentative conclusions, were to serve as encouragements. The 'Ladder of the Intellect', which was 'nothing more than an application of the second part of the Work in detail', was to include 'models, by which the entire process of the mind and the whole fabric and ordering of discovery from beginning to end ... should be set, as it were, before the eyes'. The fifth part, being certain

'Forerunners', were 'things as I have myself discovered, proved or added; not however by the true rules and prescriptions of interpretation, but by the common use of the intellect in inquiring and discovering'. Finally, the sixth part, 'Practical Knowledge', Bacon confessed to be 'both above my strength and beyond my hopes', although he believed he had made 'a beginning of the work' and hoped that others would be inspired to continue the labour.

This, then, was the sixfold plan for the *Instauratio magna*, the great restoration or renewal. Yet it is principally the second and third parts, which concern the novel methods Bacon proposed for his natural philosophy, which interest us here. It is appropriate to speak of 'methods' rather than 'method'; although Bacon asserted that his reformed natural philosophy had a novel form of 'induction', we should not be distracted by this word. ('For the induction of which the logicians speak, which proceeds by simple enumeration, is a puerile thing.')[19] The methodological novelty, so to speak, lay not in any form of *logic* but in the intellectual discipline and the institutional arrangements which Bacon demanded for the production of natural knowledge.

Bacon was convinced of the need to 'purge the floors of men's understandings' if his plan for a new natural philosophy was to succeed. In *Advancement of Learning* and (most famously) in Book I of *Novum Organum*, Bacon expounded his beliefs about the mental impediments to true learning under the heading: 'The Idols of the Mind'. Yet Bacon's first announcement of this problem is to be found in a tract of 1589, and it is only by recalling what Bacon said then, and why he said it, that we can understand the full extent of what he felt the 'Idols' put at risk.

Bacon had written 'An Advertisement Concerning the Controversies of the Church' at the height of the Martin Marprelate scandal, and while a trusted member of Secretary Walsingham's security and intelligence service. By that time he was convinced that problems of knowledge were properly among the statesman's concerns, and the 'Advertisement' reflected his anxieties about political and social instability more than it did any worries about doctrinal controversy *per se*. He argued that a major cause of controversies 'is the nature and humour of some men'; there exist those who seek an inward authority over men's minds, and an even greater number who are easily led. Thus he concluded: 'Few follow

Things in themselves, more the names of Things, and most the names of their masters.'[20] This conviction was central to Bacon's explicit accounts of the 'Idols', which first appeared in 'Valerius Terminus' (*c.* 1602) and which were elaborated repeatedly thereafter. In 'Valerius Terminus', Bacon spoke of the human mind as 'an enchanted glass', declaring there were

> four Idols or false appearances of several and distinct sorts, every sort comprehending many subdivisions: the first sort, I call idols of the *Nation or Tribe*; the second, idols of the *Palace*; the third, idols of the *Cave*; and the fourth, idols of the *Theatre*.[21]

One of the effects of the Idols was 'the great error of inquiring [seeking] knowledge in Anticipations', which were 'the voluntary collections that the mind maketh of knowledge; which is every man's reason'.[22] Here Bacon deployed one of his principal epithets, 'voluntary', and since 'voluntaries' were politically dangerous, Bacon is asserting that to seek knowledge in the 'voluntary collections' (i.e., 'every man's reason') is not simply an error but one with political consequences as well. When meditating upon human intellectual failings, Bacon typically did so from the perspective of the statesman or 'lawgiver' he wished to be. The operation of the Idols caused error in our learning and could provoke sects and then instability in the state. The existence of the Idols was a political issue for Bacon for another, greater reason, as well. Because they impede our ability to learn the rules of nature, the Idols prevent us from learning the true rules of policy. Bacon's reflections upon the words of the author of *Ecclesiastes* are highly revealing:

> 'God hath made all things beautiful, or decent, in the true return of their seasons: Also he hath placed the world in man's heart, yet cannot man find out the work which God worketh from the beginning to the end': declaring not obscurely, that God hath framed the mind of man as a mirror or glass, capable of the image of the universal world, and joyful to receive the impression thereof, as the eye joyeth to receive light, and not only delighted in beholding the variety of things and vicissitudes of times, but raised also *to find out and discern the ordinances and decrees, which throughout all those changes are infallibly observed.* And although he doth insinuate that the supreme or summary law of nature, which he calleth, 'the work which God worketh from the beginning to the end', is not possible to be found out by man; yet that doth not derogate from the capacity of

> the mind, but may be referred to the impediments ... whereunto the condition of man is subject.[23]

'To find out and discern the ordinances and decrees' by which God governs the world of things was indeed possible, but the preparatory 'purging of the floor of men's understandings' required a sustained and three-pronged attack: 'the refutation of the Philosophies; the refutation of the Demonstrations; and the refutation of the Natural Human Reason'.[24] Bacon believed that the effect of these 'refutations' would be 'the true and legitimate humiliation of the human spirit',[25] and he announced that until such time as the mind of a prospective investigator into natural knowledge was successfully 'purged',

> let him correct by seasonable patience and due delay the depraved and deep-rooted habits of his mind; and when all this is done and he has begun to be his own master, let him (if he will) use his own judgement.[26]

In other words, Bacon was arguing that due to the 'Idols of the Mind', any who asserted themselves to be their own master (i.e., 'voluntaries') could not discover truly important natural knowledge, namely, God's 'ordnances and decrees' for natural things. Furthermore, until such time as one had recovered, through the 'true and legitimate humiliation of the spirit', a prelapsarian 'nakedness of mind',[27] one could not use one's own judgement, but must accept the discipline of Bacon's 'New Instrument' in its place.

> The universe to the eye of the human understanding is framed like a labyrinth; presenting as it does on every side so many ambiguities of way, such deceitful resemblances of objects and signs, natures so irregular in their lines, and so knotted and entangled. And then the way is still to be made by the uncertain light of the sense, sometimes shining out, sometimes clouded over, through the woods of experience and particulars; while those who offer themselves as guides are ... themselves also puzzled, and increase the number of errors and wanderers. In circumstances so difficult neither the natural force of man's judgement nor even any accidental felicity offers any chance of success. No excellence of wit, no repetition of chance experiments, can overcome such difficulties as these. Our steps must be guided by a clue, and the whole way from the very first perception of the senses must be laid out upon a sure plan.[28]

Bacon's sure plan for the 'interpretation of Nature' was to begin with information derived from the senses and then, by means of a 'true induction', discover 'axioms' and (finally) the 'summary laws' which governed nature. Because of the damaging effects of the Idols on the working of the mind, Bacon insisted that men must rely upon his 'Instrument' rather than upon their own judgement:

> Now my method, though hard to practise, is easy to explain; and it is this. I propose to establish stages of certainty. The sense, guarded by a process of correction, I retain. But the work of the mind which follows the sense, I for the most part reject; and instead of it, I reveal and make open a new and certain way for the mind to proceed in, starting directly from the very perceptions of the senses ... The mind itself [must] be from the very outset not left to take its own course, but always guided; and the business be done as if by machinery.[29]

Crucial to the plan was the careful composition of 'Natural and Experimental Histories'. These were intended to vet the information of the senses, and to counteract the distortions and wishful thinking which men unwittingly made in striving to discover the principles of nature:

> Now my directions for the interpretation of nature embrace two kinds; the one how to produce and summon forth axioms from experience; the other how to deduce and derive new experiments from axioms. The former again is divided into three ministrations; a ministration to the sense, a ministration to the memory, and a ministration to the mind or reason. For first of all we must prepare a Natural and Experimental History, sufficient and good; and this is the foundation of all.[30]

These histories, which Bacon called 'Primary Histories', were to be the product of collecting information about the natural world as it was, and also information about the natural world derived by 'artifice', that is to say, by 'experiments'. Bacon's natural philosophy

> neither relies solely or chiefly on the powers of the mind, nor does it take the matter which it gathers from natural history and mechanical experiments and lay it up in the memory whole, as it finds it; but lays it up in the Intellect *altered and prepared*. Therefore, from a closer and purer league between these two faculties, the experiental and the rational (such as has never yet been made), much may be hoped.[31]

These histories would be scrutinised repeatedly by means of the 'true induction' until general, 'higher' axioms were produced:

> Now my plan is to proceed regularly and gradually from one axiom to another, so that the most general are not reached until last: but then when you do come to them you find them to be not vague but well-defined notions, and such as Nature would really recognise as her first principles and such as are at the heart of things.[32]

The axioms would approximate with increasing accuracy the 'true Forms' of nature, that is to say, God's 'summary Laws' for the natural world. Only after this knowledge had been obtained might men turn effectively to 'Works':

> I hold it enough to have constructed the machine, though I may not succeed in setting it on work. Nay with the same candour I profess and declare, that the Interpretation of Nature, rightly conducted, ought in the first steps of the ascent, until a certain stage of Generals be reached, to be kept clear of all application to Works.[33]

Let us look in more detail at Bacon's 'machine' for the interpretation of nature: first, the construction of 'Natural and Experimental Histories', secondly, the scrutiny of such histories and the discovery of 'Forms' and, finally, the nature of the 'Forms'.

Bacon insisted that natural history should be 'of three sorts: of nature in course; of nature erring or varying; and of nature altered or wrought'.[34] Others, he claimed, thought that dealing with the first two sorts was sufficient, but because natural history properly 'recalls the deeds and actions of Nature' as a whole,[35] natural history must also include instances of 'nature altered or wrought', that is to say, a 'History of Arts, which I also call Mechanical and Experimental history'.

> And I am the rather induced to set down the History of Arts as a species of natural history, because it is the fashion to talk as if art was something different from nature, so that artificial things should be separated from natural ones, as differing totally in kind; whence it comes that most writers of natural history think it enough to bring together a history of animals or plants or minerals, omitting the experiments of mechanical arts (which by far are of the greatest influence for philosophy); and not only that, but another and more subtle evil finds its way into men's thoughts; that of looking upon art merely as a kind of addition to nature; which has power enough to

finish what nature has begun or correct her when deviating, but no power to make radical changes, and shake her in the foundations; an opinion which has brought much desperation into human concerns.[36]

Because of this, and because 'the nature of things shows itself to a greater extent under the vexations of art than in its natural freedom',[37] the 'Mechanical or Experimental Arts' were the crucial part of a successful natural history:

> Among the parts of history ... the history of arts is of greatest use, because it exhibits things in motion, and leads more directly to practice. Moreover it removes the mask and veil from natural objects, which are commonly hidden and obscured under the variety of shapes and external appearance. Finally, the vexations of art are certainly as the bonds and manacles of Proteus, which show the deepest strainings and efforts of matter. For bodies will not be destroyed or annihilated; instead, they will turn themselves into various forms. Therefore, upon this history, mechanical and illiberal as it may seem (all arrogance and fastidiousness abandoned) the greatest diligence must be bestowed.[38]

Bacon offered explicit instructions about how to prepare a successful 'primary History'. In the first place, much that one usually found in natural histories he condemned as 'matters superfluous':

> First then, away with antiquities, and citations or testimonies of authors, also away with disputes and controversies and differing opinions ... Also let all those things which are admitted be themselves set down briefly and concisely ... Secondly, that extravagance of natural histories in descriptions and pictures of species, and the curious variety of same, is not much to the purpose ... Thirdly, all superstitious stories ... and experiments of ceremonial magic should be altogether rejected ... It is always to be remembered that this which we are now about is only a granary and storeroom of matters ... only to be entered as occasion requires, when anything is wanted for the work of the *Interpreter*, which follows it.[39]

All the pieces of information that the historian deemed worthy of being entered into this 'storeroom' were to be carefully considered as well:

> With regard to the credit of the things which are to be received into the history, necessarily they be either certainly true, doubtful whether true, or condemned as untrue. Things of the first kind should be

> announced simply, the second with a note, namely, by the words 'it is reported', 'they relate', 'I have heard from a person of credit', and the like.[40]

The collector of a 'primary history' was to indicate the degree of credibility he associated with any item he reported. Bacon thought it advantageous as well that the historian should add certain further comments upon the collected materials:

> Firstly, questions (not about causes but about facts) should be added in order to provoke and stimulate further inquiry ... Secondly, in any new and more subtle experiment the manner in which the experiment was conducted should be added, that men may be free to judge whether the information obtained from that experiment be trustworthy or deceiving ... Thirdly, if in any statement there be anything doubtful or dissatisfying, by no means would I have it suppressed or passed in silence, but plainly and perspicuously set down by way of note or admonition. For I want this primary history to be drawn up most religiously, as if every particular were stated upon oath.[41]

This discipline in reporting was crucial in preparing successful 'primary history', for it helped to minimise the distorting effects of 'unguided Reason':

> Not only is a greater abundance of experiments to be sought for and procured, and that also of a different kind from those already made; an entirely different method, order, and process for carrying on and advancing Experience must be also introduced. For Experience, when it wanders in its own track, is ... merely groping, and confounds men rather than instructs them. But when it shall proceed in accordance with a fixed law, in a regular and uninterrupted order, then may better things be hoped of the sciences.[42]

The experiments which the historian himself conducted and then included in his primary history were expected to repair and 'advance' men's wayward 'experiential faculty':

> For the subtlety of experiments is far greater than that of the sense itself, even when assisted by exquisite instruments; such experiments, I mean, as are skilfully and artfully devised for the express purpose of investigating the point in question. Therefore, to the immediate and proper perception of the sense I do not attribute much; but I suggest

> that the sense shall judge of the experiment and the experiment shall judge of the thing. And thus I conceive that I perform the office of a high-priest of the sense (from which all knowledge in nature must be sought, unless men are to go mad) and a not unskilled interpreter of its oracles.[43]

These experiments would be prompted by those collected items of information which the historian noted as dubious, and by the queries he appended to a particular report. The queries, designed 'to provoke and stimulate further inquiry', would allow the framing of a 'point in question' which would be addressed by a 'more subtle experiment'.

Bacon used the single term 'experiment' to describe very different sorts of activity by the natural historian: the passive reporting both of observed craft practices and techniques, and of particular inquiries conducted by other men; the 'artificial' investigations he carried out himself; and any subsequent, 'more subtle', investigations. For example, Bacon's collection of 'experiments' entitled *Sylva Sylvarum, or a Natural History* displays him engaged in all of these activities. Some 'experiments', however, were designed to do more than swell a natural history:

> Those experiments be not only esteemed which have an immediate and present use, but those principally which are of most universal consequence for invention of other experiments, and *those which give most light to the discovery of causes.*[44]

'The discovery of causes', of course, was the work of the 'Interpreter of Nature' who examined a natural history, yet 'causes' themselves were of differing value, as Bacon made clear in *Of the Proficience and Advancement of Learning*:

> Natural history describeth the variety of things; physic the causes, but variable and respective causes; and metaphysic the fixed and constant causes.[45]

The lesser causes, which Bacon ascribed to the domain of 'physic', were the doorway to the greater ones (otherwise described as the 'forms of Nature'), the domain of true 'metaphysic':

> Metaphysic is most excellent in two respects: the one, because it is the duty and virtue of all knowledge to abridge the infinity of individual experience, as much as the conception of truth will permit,

> and to remedy the complaint of *vita brevis, ars longa*; which is performed by uniting the notions and conceptions of sciences. For knowledges are as pyramides, whereof history is the basis. So of natural philosophy the basis is natural history; the stage next the basis is physic; the stage next the vertical point is metaphysic. As for the vertical point, *opus quod operatur Deus a principio usque ad finem*, the summary law of nature, we know not whether man's inquiry can attain unto it. But these three be the true stages of knowledge and ... as the three acclamations, *Sancte, sancte, sancte!* holy in the description or dilatation of his works; holy in the connexion or concatenation of them; and holy in the union of them in a perpetual and uniform law ... So then always that knowledge is worthiest which is charged with least multiplicity, which appeareth to be metaphysic; as that which considereth the simple forms and degrees and co-ordinations whereof make all this variety. The second respect, which valueth and commendeth this part of metaphysic, is that it doth enfranchise the power of man unto the greatest liberty and possibility of works and effects. For physic carrieth men in narrow and restrained ways, subject to many accidents of impediments, imitating the ordinary flexous courses of nature ... For physical causes give light to new invention in *simili materia*. But whosoever knoweth any form, knoweth the utmost possibility of superinducing that nature upon any variety of matter.[46]

In 'Cogitata et visa' (*c.* 1607), Bacon first described how the 'Interpreter' was to scale the 'pyramid' of natural philosophy. In the first place:

> He thought that a body and mass of particulars both from their number, kind, and certainty or subtlety sufficient for information, might be collected and stored up, both from natural history and mechanical experiments ... He thought that this mass should be reduced and digested into Tables and regular order, that the Intellect may be able to act upon it.[47]

The information of a primary natural and experimental history, then, was to be 'reduced and digested' into 'Tables', and it was by means of these that all subsequent stages would be reached. Here Bacon offered two important warnings:

> He thought that we must not suddenly pass from the particulars referred into tables, to an inquisition of new particulars (which nevertheless is itself a useful thing and like a kind of learned

> experience), but that we should first ascend to general and common comprehensions.

To descend now from the tables 'to the inquisition of new particulars' was no part of the plan, but an activity he termed 'learned experience' (*experientia literata*) and which he later was to describe as 'the mere transferring of the experiments of one art to others' and to dismiss, declaring 'no great matter ... can be hoped from that'.[48] Yet this useful, but diversionary, manœuvre was not the only temptation to be avoided by the Interpreter:

> He thought that the natural but mis-guided motion and impulse of the Intellect to jump from particulars to the highest and most general comprehensions (such as are called the principles of arts and things), and to supply the rest by descending through the middle ones, must be altogether suppressed.

These warnings given, Bacon then gave positive instructions on how to proceed:

> The nearest comprehension must be drawn out and discovered first, and then the middle ones, and we must ascend the true ladder by successive steps ... He thought that such a form of induction should be devised as should from particular instances finish in general terms, so long as it can be shown that a contradictory instance cannot be found, lest by chance we pronounce from a smaller number than is fitting ... He saw that that comprehension only should be assented to and admitted which was not made and fitted to the measure of the particulars from which it was derived, but which was rather more ample and extensive and confirmed its amplitude or latitude by the designation of new particulars as a sort of guarantee, lest we should stick at what is already known, or perchance in too wide an embrace clutch at shadows and abstract forms.[49]

In other words, the Interpreter's 'first comprehension' (i.e., his conjecture as to a 'lesser axiom') was to be 'drawn out and discovered' from the 'Tables' and if judged to be certain, then it and other such approved 'comprehensions' would supply materials for the discovery of 'middle comprehensions' (and 'middle axioms'). The process of judging a 'first comprehension' to be certain was of two parts. The initial part involved searching the history for a 'contradictory instance'. If one was found, then (presumably) an alternative 'first comprehension' would have to be conceived. If none was to be found, then the Interpreter was to ensure the 'comprehension' was framed in terms 'more ample' than would include merely 'the particulars from which it was derived'; it was to hold valid for further particulars as well. This act of extension was to be supported by means of new experiments (i.e., by discovering new 'particular instances') in order to test the 'comprehension'

> with regard to the first notions of the intellect; there is not one of these taken by the intellect when left to go its own way, but I hold it for suspect and no way authoritative until it has submitted to a new trial and a fresh judgement has been pronounced.[50]

This elaborate process of affirming the 'first notions', or 'comprehensions', Bacon often referred to as a process of 'exclusion and rejection' for the establishment of axioms:

> In establishing axioms, a form of induction other than has been employed as yet must be devised; and it must be used not for proving and discovering first principles (as they are called) only, but also the lesser axioms, and the middle, and indeed all ... the induction which ought to be available for the discovery and demonstration of sciences and arts, must analyse nature by proper rejections and exclusions; and then, after a sufficient number of negatives, come to a conclusion on the affirmative instances.[51]

The 'tables and arrangements of instances' were designed to facilitate this process of 'rejection and exclusion'. Instances drawn from a natural history were to be challenged, as it were, by other, seemingly contradictory, instances which were to be arrayed against them. When the Interpreter examined those 'affirmative instances' which were not challenged, he was then able to make a 'first comprehension'.

In Book II of *Novum Organum*, Bacon described how to begin an investigation into the 'form of heat'. He first listed a series of 'affirmative instances' concerning heat which he termed a 'table of essence and presence' (e.g., the rays of the sun, fiery meteors, flames – all being things which were hot). This was followed by a series of negative instances, arranged as a 'table of deviation, or of absence in proximity' (e.g., the rays of the moon and stars; comets, assumed to be a sort of meteor, do not exert any effect in increasing the heat of seasons; however, no negative was found of flames, all instances of which are hot). On the basis of the surviving 'affirmative instances' in the first table, a third table ('of degrees') was then established, listing those earthly 'substances which are hot actually, and to the touch', and estimating their relative 'intensities and degrees' of hotness.

> The work and office of these three tables I call the Presentation of Instances by the Intellect. This presentation having been made, Induction itself must be set at work; for the problem is, upon the presentation of the instances, all and each, to find such a nature as is always present or absent with the given nature [Heat], and always increases and decreases with it; and which is ... a particular case of a more general nature ... The first work therefore of true Induction (as far as regards the discovery of Forms) is the Rejection or Exclusion of the singular natures which are not found in another instance where the given nature is present, or are found in some instance where the given nature is absent, or are found to increase in some instance when the given nature decreases, or to decrease when the given nature increases. Then truly after the Rejection and Exclusion has been duly made, there will remain at the bottom, all light opinions vanishing into smoke, a Form affirmative, solid and true and well-defined.[52]

This 'first work', a process of 'exclusion or rejection of ... several natures', Bacon used to delimit 'a conjecture as to the Form'. For example:

> By a presentation of ignited iron and the flame of spirit of wine (of which ignited iron has more heat and less brightness, while the flame of spirit of wine has more brightness and less heat) *reject* ... light and brightness.[53]

In other words, 'light' and 'brightness' vary greatly in different instances of 'heat' and therefore, he declared, they 'do not belong to the Form of Heat'. They are, in fact, 'singular natures' in their own right. By such means, the Interpreter was gradually to consolidate a definition of the 'form' of heat.

In addition to the procedures outlined here, Bacon offered a series of supplementary techniques for speeding up and bolstering the certainty of the process. There were nine of these techniques, the names of which Bacon listed,[54] but only the first of which he described: namely, the 'prerogative instances', of which he listed twenty-seven sorts.[55] Nine of these were to be used to frame additional natural histories.[56] The rest 'need not be inquired into until we come to make tables of comparison':

> But, the choosing of the pre-eminent instances in every single genus (which ought principally and most diligently to be sought and, as it were, hunted out) is to be demanded from the prerogatives of instances.[57]

These eighteen 'prerogative instances' were to help in the 'process of exclusions' by which the Interpreter's 'near comprehensions' were first arrived at.[58]

The Interpreter's next task, namely 'ascending' from the 'lesser axiom' (i.e., the verified 'first comprehension') to a 'middle axiom' was to be done by means of what Bacon called (rather confusingly) the 'beginning of interpretation, or the first vintage'. The 'first vintage' consisted of two phases: the forming of a conjecture as to the 'middle axiom', and then a process of verifying and refining that claim.

Those 'affirmative instances' which survived the 'process of exclusions' were now re-examined by the Interpreter who, now satisfied as to a 'lesser axiom' based upon those instances, hoped to make a conjecture as to a 'middle axiom':

> From all and each of the instances, the nature of which Heat is a limitation appears to be Motion ... When I say of Motion that it is as the genus of which Heat is a species, I would be understood to mean not that heat generates motion, or that motion generates heat (though both are true in certain cases), but that Heat itself (or, its quiddity), is Motion and nothing else.[59]

'Heat', then, is a form in itself, but the Interpreter's 'middle conjecture' is that 'heat' is subordinate to 'motion', a 'higher and more general nature'. Very importantly, the Interpreter is no longer engaged in new 'experiments' (the producing of new particular instances); he is relying now on those discovered either in building up natural histories or in arriving at his 'lesser axiom' (in this case, 'heat'). He has such confidence in his earlier labours that he has abandoned the 'experiential faculty' in favour of the 'rational faculty', and is constructing a hierarchical taxonomy of the genera and species of 'forms'.

In order to refine his conjecture about the 'middle axiom', Bacon now listed a set of 'differences' or specifications (and the 'affirmative instances' in which these *differentiae* were displayed) by which 'motion' was delimited into the species 'heat'. In other words, the 'differences' were the defining characteristics of a certain subdivision (a species) of 'motion' (a genus):

> The first Difference, therefore, is this. Heat is an Expansive motion, whereby a body strives to dilate and stretch itself to a larger sphere or dimension than it previously had occupied. This Difference is most observable in flame, where the smoke or thick vapour manifestly dilates and expands itself into flame. Also, it is shown in all boiling liquid, which manifestly swells, rises and emits bubbles; and carries on the process of self-expansion, until it turns into a body far more extended and dilated than was the liquid itself, namely, into vapour, smoke, or air ... The second Difference is a modification of the former; namely that heat is a motion expansive or towards the circumference, but with this rule, that the body has at the same time a motion upwards ... The third Difference is this; that Heat is a motion of expansion, not uniformly of the whole body together, but in the smaller particles; and at the same time checked, repelled, and beaten back, so that the body acquires a motion alternative, perpetually quivering, striving and struggling, and irritated by repercussion, whence springs the fury of fire and heat ... The fourth Difference is a modification of the last; it is, that that motion of stimulation or penetration must be somewhat rapid and not sluggish, and must be performed by particles, minute indeed ...[60]

Armed with this array of specific differences within the genus *motion*, Bacon could announce a definition of heat:

> From this First Vintage, it follows that the Form or true definition of Heat (heat, that is, in relation to the universe, not simply in relation to the senses) is, in a few words, as follows: *Heat is a Motion, expansive, restrained, and acting in its strife upon the smaller particles of bodies*. But the expansion is thus modified: *while it expands all ways, it has at the same time an inclination upwards*. And the struggle in the particles is modified also; *it is not sluggish, but hurried and with violence*.[61]

We can summarise the process by which Bacon hoped to discover 'Forms'. To begin, collect natural and experimental histories of 'particular instances' (from books, from men of repute, and from one's own experiments). Decide upon a 'nature' to be investigated (e.g., heat).[62] Retrieve from the natural histories all those instances that appear to have bearing upon this investigation. Conduct a 'process of exclusion and rejection', by means of the first two 'tables of comparison', eliminating those instances that have counter-instances. Arrange the remaining affirmative instances into a third table which arrays them according to their greater or lesser manifestation of the nature under investigation. Examine this table, and make an informed conjecture as to the 'lesser form' (i.e., the definition of the nature that all these instances share). Test this conjecture by re-examining the histories for any instance that might contradict it. If the conjecture 'passes' this test, then frame it in terms more liberal than is strictly justified by the known instances themselves; thus, the conjecture will not merely recapitulate the instances collected. Design new experiments to test the extension of the conjecture; these, in effect, produce new, and pertinent, instances, and (if they do not invalidate the extension), the Interpreter can be confident he has arrived successfully at a 'lesser axiom'. Ascend from lesser to middle axioms by the process known as the 'first vintage', which is concerned with species and genera and is of two phases. First, make a conjecture about the middle axiom by re-examining the affirmative instances which survived the earlier 'process of exclusions'. This is a conjecture about the genus to which the lesser form belongs. Secondly, verify this middle conjecture by listing the *differentiae* that define the lesser form as a subordinate of the conjectured middling form.

Bacon does not tell us how to proceed further. Presumably this was to follow in the promised discussion of the 'ascending and descending scale of axioms'.[63] Yet judging from his general remarks about the 'instrument', we can presume that the procedure of interpretation (i.e., *not* the compiling of natural histories) was to be repeated for other 'given natures'. In this way a clear picture was to be built up both of the species ('lesser axioms' or lesser forms) which lie within a genus ('middle axiom') and also of a wide array of genera. As for the discovery of even higher genera, and then of the 'summary laws of nature', we have no way of knowing what Bacon had in mind. Perhaps it was to be achieved by the 'rational faculty' engaging yet again in the method of specification described for discovering middle axioms. Nevertheless, what Bacon has described is perfectly consistent with his remarks about the 'pyramid' of natural history, 'physic' and 'metaphysic'. 'Heat', surely, is a discovery in 'physic', and 'motion' a discovery in 'metaphysic'. It seems perfectly clear as well that the 'new instrument' and the discovery of forms is *not* to be described as a process of 'induction', as logicians commonly understand the term.

Bacon's plan for the interpretation of nature was not a private method but a public institution. Ever since his earliest pronouncements in the 1590s, Bacon had demanded that his plans would require many helping hands if it was to succeed. He imagined it as a new department of state, a 'royal work', with a royal governor. The enigmatic motto of *Instauratio magna*'s frontispiece – the prophetic verse of Daniel 12:4 (*multi pertransibunt & augebitur scientia*) – was easily read as licensing 'voluntaries', but it was written by a Lord Chancellor, whose ambition clearly was that 'many shall go to and fro' – under his direction – 'and knowledge shall be increased'. Bacon was at great pains to point out that the collection of natural histories in particular was a vast enterprise which demanded human and material resources on a scale beyond the capacities of individuals:

> For a history of this kind ... is a thing of very great size, and cannot be executed without great labour and expense; requiring as it does many people to help, and being (as I have said elsewhere) a kind of royal work ... one must employ factors and merchants to go everywhere ...[64]

This he made perfectly clear in 'Solomon's House', a vast establishment of buildings and employees, and even the elite brethren were to be subject to a strict hierarchy of command and a careful division of labour. Moreover, among the thirty-six brethren, there were only three Interpreters, and this is perfectly consistent with his description of how the 'new instrument' was to function. The labour of collecting natural histories, which included much of the activity of conducting new 'experiments', was all preparatory to the work of the Interpreter himself.

Who was the 'Interpreter of Nature'? We can approach this by looking at some of Bacon's general remarks about the nature of forms. As we have seen, there was a hierarchy of forms, lesser ones governing 'simple natures' (such as heat), and greater ones governing them. 'Governing' is an apt description of what the forms do, since they are 'laws':

> When I speak of Forms, I mean nothing else than those laws and limits of pure act, which govern and constitute any simple nature, as heat, light, weight, in every kind of matter and susceptible subject. Therefore, the Form of Heat or the Form of Light is the same thing as the Law of Heat or the Law of Light.[65]

What Bacon presents as the work of the 'Interpreter of Nature', then, is a *scientia* devoted to discovering the hierarchy of laws, of governance, over the 'simple natures' found in natural things:

> In nature nothing really exists beside individual bodies, performing pure individual acts according to a law; yet in talking about this law itself, and the inquiry, discovery, and explanation of it, is the foundation of knowledge as well as of operation.[66]

Moreover, Bacon imagined the laws of nature to be in the shape of a hidden code with subordinate sections:

> It is this law, and its sections and clauses [*paragraphos*], that I mean by the name of *Forms*; a term which I adopt especially since it has grown into use and become familiar.[67]

The 'sections and clauses' were as the lesser and middling forms, and the highest forms, the 'summary laws of nature', were the principles by which God governed the world. Thus Bacon could announce that 'the laws of heaven and earth ... are the subject of philosophy'.[68] After all, in both the ancient Latin and the Elizabethan English, an *interpres* or 'interpreter', was an official expounder of the laws. The master of the 'new instrument', then, was a new sort of royal judge, and the *Instauratio magna* justifies and describes his work.

Law and natural philosophy

We have claimed that Bacon's reformed natural philosophy was modelled on his reforms of the law. As we saw, English law should be understood not merely as a body of prohibitions, technical instruments and court procedures but, more importantly, as a royal structure for the governance of England. Bacon desired massive reforms of the existing English science of the law in order to render it more agreeable to the will of its master, the king. So, too, with his natural philosophy: it was a department of state, and his new instrument was an institutional structure for the centralised governance of inquiry into the natural world. His motives for reforming natural philosophy and for reforming English law were the same: the desire to create the conditions for empire. Yet the parallels between Bacon's reformed natural philosophy and his reformed science of law are more striking than a common purpose alone. Parallels exist between the conduct of a lawsuit and that of an experiment, and between the science of the law and the science of nature.

Although he repeatedly claimed that 'experiments' were a crucial part of his reformed natural philosophy, Bacon never provided a detailed discussion of how a particular new experiment or investigation was to be conducted. Yet he made enough remarks in passing to show that he envisioned an 'experiment' to be closely similar to a lawsuit and its trial in a court room.

Both in lawsuits and in Bacon's natural philosophy, the first and crucial stage was the 'establishment of facts'. This required the gathering of testimony. In law, the principal means of gathering information was the use of 'schedules of interrogatories', and Bacon is explicit that he means to do the same in natural philosophy: 'I mean (adopting the example from civil causes) ... to examine Nature herself ... upon interrogatories [*super articulos*].'[69]

In Bacon's natural philosophy, 'interrogatories' were to be used in the preliminary stage, namely, the gathering of 'experiments' for the natural histories, where the greatest care needed to be taken to ensure that the information collected was completely certain. Bacon's procedure was very much that of the Star Chamber, where everyone, including the principals, was examined upon oath; he declared: 'I want this primary history to be drawn up most religiously, as if every particular were stated upon oath.'[70] Schedules of interrogatories were drawn up and vetted with extreme care before they were deployed, and the responses of those examined were recorded with equal care. In some situations, such as sedition investigations, the answers to a schedule would be scrutinised and used to help frame subsequent schedules put to the prisoner or witness. Bacon advocated a similar process in building up a primary history:

> Firstly, questions (not about causes but about facts) should be added in order to provoke and stimulate further inquiry ... Secondly, in any new and more subtle experiment, the manner in which the experiment was conducted should be added, that men may be free to judge whether the information obtained from that experiment be trustworthy or deceiving ... Thirdly, if in any statement there be anything doubtful or dissatisfying, by no means would I have it suppressed or passed in silence, but plainly and perspicuously set down by way of note or admonition.[71]

Such queries would allow the framing of a 'point in question', which would be addressed by a 'more subtle experiment'.[72] Whenever judicial torture was employed, its purpose was to force responses to interrogatories from the prisoner, and we can recall Bacon's insistence that nature best revealed herself when 'vexed', and 'tortured'.[73] Nor should it surprise us now that a diligent and highly motivated security officer such as Bacon should speak of removing 'the mask and veil from natural objects which are commonly hidden and obscured under the variety of shapes and external appearance'.[74] The legal process of interrogatories and depositions was designed to discover what the 'deed' or 'fact' actually was; similarly, in Bacon's natural philosophy, natural things are 'vexed' by 'artificial experiments' in order to reveal their 'motions'.

The testimony of different witnesses was taken by judge and jury to be of differing credibility. In particular, credibility differed according to the status of the witness, and Bacon incorporated this consideration into his directions for making natural histories. For any information which was 'doubtful whether true', the historian should be careful to note whether he had the report 'from a person of credit' or otherwise.[75]

The 'establishment of facts' required more than the careful gathering of testimony, since a 'fact' was a construction of the court room. Not only the gathering of testimony, but the dynamics of the court trial as well had its counterpart in Bacon's natural philosophy. It was in the process of 'well pleading' that the competing barristers finally agreed upon what constituted the facts of a case. The formation of 'tables' in Bacon's description of his new instrument functioned in this adversarial manner. Admittedly, the natural history, rather than two independent sources, provided the

Interpreter with both tables of instances – but in Bacon's proposed revision of the law reports, the lawyers, too, would have had to rely upon one set of reports. We may reflect here upon Bacon's use of a word which emphasised the parallel he envisioned with the court room. The Latin phrase earlier translated as 'tables of the presentation of instances to the intellect' contains a legal term. *Comparentia* ('presentation') refers to the appearance in a court room of a party to a cause.[76] 'Instances' are eventually summoned to appear in the 'court room', as it were, of the Interpreter, who is an expositor of the laws, namely, a judge.

In cases that turned upon 'issues of law', the barristers engaged in a process of argument by analogy with older cases in order to discover the hidden law. This process of 'case-law reasoning' is extremely important for Bacon's natural philosophy. A barrister would claim that underlying the decisions in a specific array of older cases was a particular legal principle which governed all of those decisions, and he desired that the judge invoke this in deciding the present case. His opponent would offer a different array of cases and petition for judgement based upon a different principle of law. The judge would have to assess the strength of these claims. Given his 'case' (a 'given nature'), the Interpreter would similarly have to adjudicate between two competing tables of instances, and he did this by a 'table of degrees' in which the relative worth of the instances was measured. On the basis of this, he would form a private conjecture as to an 'axiom'. Both the judge and Bacon's Interpreter were seeking to determine which legal principle (or natural species) it is which is manifest in the case (or given nature) under review. During cases about 'issues of law', a judge was presented with the opportunity and appropriate materials for discovering the hidden law; similarly, Bacon declared there were certain 'experiments' which 'give most light to the discovery of causes'.[77] It was on the basis of his learned interpretation of the matter thus presented that the Interpreter eventually would determine a particular case or given nature with a judgement or 'axiom'. This decision then was recorded for future reference.

The process of legal reasoning was described by ancient orators and jurists, notably by Cicero and Quintilian, as 'induction'. This use of the term is different from the way it was typically used by natural philosophers and logicians from Aristotle onwards. We can recall that Bacon regarded 'the induction of which the logicians speak' as 'simple enumeration', 'a puerile thing'. It is entirely likely that the Lord Chancellor had this legal usage in mind when he declared the interpretation of nature was dependent upon 'a true induction'.[78]

Bacon assumed there was a similarity between the nature of the common law and the laws of nature. He believed that much of the common law was hidden, that it was rational, and that men 'learned in the law' could discover it. The same was true of God's natural world. The principles of nature were hidden, yet they were capable of being discovered by the learned and licensed investigator. Not surprisingly, the science of discovering the law (as reformed by Bacon) and his science of discovering axioms in nature share the same structure.

Both Bacon's reformed science of the law and his natural philosophy are pursued by official investigators alone. Their work, according to Bacon, is based upon the careful compilations known as law reports and natural histories, and Bacon regarded these as similar: law reports, he said, were 'a kind of history ... of the laws'.[79] Both the royal judge and the Interpreter were to scrutinise these to discover the hidden 'axioms', 'maxims', or 'rules', and then to discover 'laws of laws' and 'higher axioms of nature'. They would sort the individual 'maxims' or 'axioms' into a hierarchy of species and genera, and this is how they gradually discovered the hidden structure of the common law or of nature. Book II of *Novum Organum*, then, describes the work of a judge.

We may recall as well the arrangements of 'Solomon's House'. Here the parallels with Bacon's reformed structure of law are striking. The brethren of Solomon's House are clearly akin to Bacon's royal judges: like the judges, they are state officers; they are distinguished by their sumptuous robes of office; they receive great deference; they go on 'circuits', and they deliver counsel and 'remedies' to the people. Elizabethan judges, moreover, spoke of themselves as 'learned brethren'. Yet the parallels are stronger than this, as two charts found in the previous chapter clearly illustrate. The first chart (see p. 138), the structure for producing knowledge in Solomon's House, mirrors the second one (p. 119), the structure which Bacon proposed for a reformed science of law.

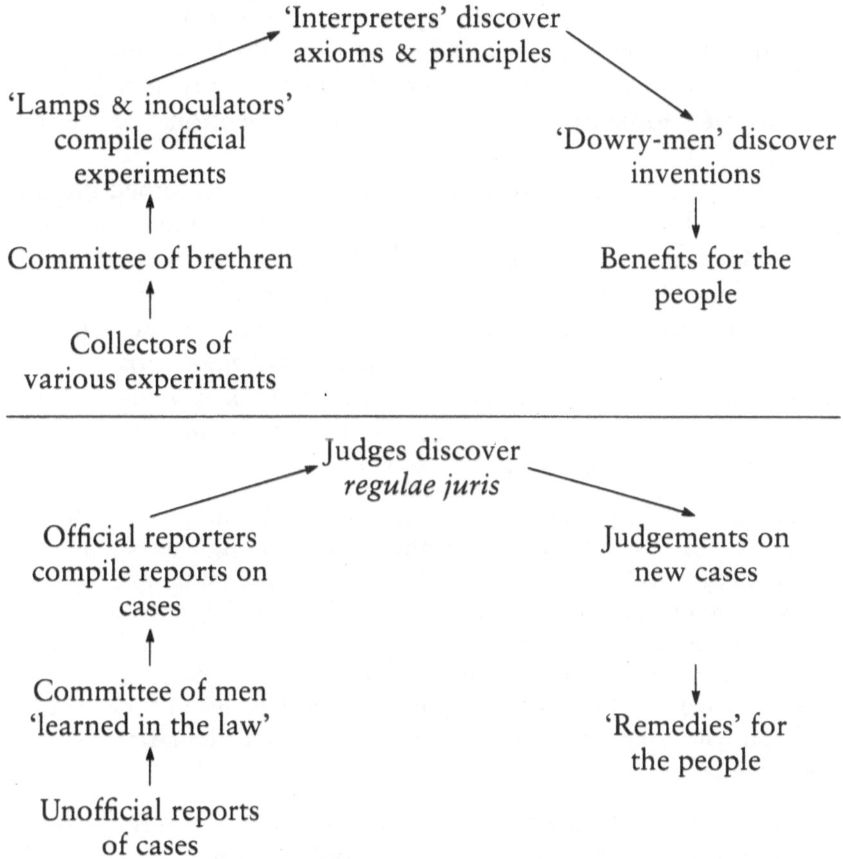

Every stage Bacon proposed for the process of acquiring knowledge about the natural world has an exact parallel in the process he proposed for acquiring knowledge about the law. On the one hand, we are to begin with the collection of 'experiments' by 'pioneers' and the like, and on the other, with the collection of cases of 'issues of law' by 'voluntaries'. These collections then pass to an official committee which decides which experiments (or law cases) are truly important, and then official collections are compiled. The process of creating natural histories is analogous to that of creating official law reports. These form the basis of all subsequent work, namely, the work of the Interpreter (who aims to discover the hidden principles of nature) and also the work of the learned Baconian judge (who aims to discover the hidden *regulae juris*).

These discoveries then are recycled into the official store of learning, and help the *interpres* discover further, and higher, principles. It is from these discoveries as well that practical benefits flow to the people in the form of practical inventions (or 'certain laws').

Bacon's reformed science of the law and his reformed natural philosophy have the same purpose, the same techniques, the same vocabulary and the same hierarchical organisation. The king's Lord Chancellor, who as a young man declared 'all knowledge to be my province', expected that his critical survey of 'all learning' (e.g., in *The Advancement of Learning* and *De augmentis scientiarum*) and his novel proposals for attaining knowledge of the natural world (e.g., in *Instauratio magna*) would lead to an 'administration of learning':

> No man hath propounded the general state of learning . . . A just story of learning, containing the antiquities and originals of knowledges and their sects, their inventions, their traditions, their diverse administrations and managings, their flourishings, their oppositions, decays, depressions, oblivions, removes, with the causes and occasions of them . . . I affirm to be wanting. The use and end of which work . . . [is] for a grave purpose, which is this in a few words, that it will make learned men wise in the use and administration of learning.[80]

It was from this statesman's point of view that Bacon spoke of his desire to master what he called 'philosophia prima'. He employed this term to describe something utterly different than did philosophers since the ancient Greeks:

> I intend Philosophia Prima, Summary Philosophy, and Metaphysic, which heretofore have been confounded as one, to be two distinct things. For the one I have made as a parent or common ancestor to all knowledge, and the other I have now brought in as a branch or descendent of Natural Science.[81]

By 'philosophia prima', Bacon was alluding to the knowledge of the 'great affinity' or 'congruity' he believed existed between 'the principles of Nature and Policy', and which he believed to have been possessed by the ancient Persian magi, the Gymnosophists and others, who had been the councillors of the greatest and imperial kings.

> My meaning touching this original or universal philosophy is thus ... *That it be a receptacle for all such profitable observations and axioms as fall not within the compass of any of the special parts of philosophy or sciences, but are more common and of a higher stage.*
>
> Now that there are many of that kind need not be doubted. For example, is not the rule, *Si inaequalibus aequalia addas, omnia erunt inaequalia*, if equals be added to unequals, the wholes will be unequal, an axiom as well of justice as of mathematics? And is there not a true coincidence between commutative and distributive justice, and arithmetical and geometrical proportion? Is not the ground, which Machiavel wisely and largely discourseth concerning governments, that the way to establish and preserve them is to reduce them *ad principia*, a rule in religion and nature as well as in civil administration? Was not the Persian Magic a reduction or correspondence of the principles and architectures of knowledge to the rules and policy of governments? ... Neither are these only similitudes, as men of narrow observation may conceive them to be, but the same footsteps of nature, treading or printing upon several subjects or matters.[82]

Because 'philosophia prima' was the learning of 'administrators of knowledge', rather than that of men learned in any individual science, Bacon desired

> an administration of knowledge in some such order and policy as the king of Spain in regard to his great dominions useth in state; who though he hath particular councils for several countries and affairs, yet hath one council of state or last resort, that receiveth the advertisements and certificates from all the rest.[83]

In his allusion to the manner by which the king of Spain governed his great empire, Bacon revealed yet again his political perspective on matters of learning, and his insistence that natural philosophy, properly governed, would help bring about an empire for his master. This insistence was presented in the boldest possible way in the frontispiece to *Instauratio magna* itself, and what more appropriate place for this central message from a statesman and Privy Councillor than at the beginning of his greatest exposition of a reformed natural philosophy?

Conclusion

Why may not that be the skull of a lawyer? Where be his quiddities now, his quillities, his cases, his *Tenures*, and his tricks?

Shakespeare, *Hamlet*, V, i, 98–100

If historians are to understand what it was that Bacon thought he was doing in writing about natural philosophy, we must accept that he always regarded himself as a statesman and a personal councillor to kings, rather than as a private person. By examining his upbringing and his legal career as an elite Crown servant, we have been able to see that his reform of natural philosophy, no less than his proposed reform of law, was always governed by his political perspective and his loyal ambition to create bureaucratic machinery with which his master could better govern and expand his kingdom. Bacon was convinced that the enhancement of the powers of the Crown was a great and necessary good and, as his loathing for men like Edward Coke illustrates, he viewed any attempts to detract from these powers as seditious. Similarly, in Scriptural matters, men should accept the leadership of official experts, namely, the ecclesiastical hierarchy; those who assumed a right to pronounce independently about religious truths were 'sectaries'. So, too, in natural philosophy: men should accept the leadership of official experts and a centralised organisation. Those who assumed otherwise were 'willful' and 'voluntaries'. No wonder that his physician, William Harvey, a celebrated voluntary (as it were), could say that Bacon wrote philosophy 'like a Lord Chancellor'.[1]

The social terrain in which the study of natural philosophy was pursued during the sixteenth century expanded from the university and cloisters into the broader and more volatile public arena: not only scholars, but a widening array of artisans and gentlemen were asserting a right to pursue natural knowledge and to make public pronouncements about its substance, proper methods and uses.[2] Such expansiveness was not universally welcomed; it was regarded in some quarters, by men valuing official mediation in natural knowledge above liberality in its pursuit, as potentially dangerous to the structure of civil society. For example, in the early 1590s when Bacon was making his first pronouncements about natural philosophy, the poet (and occasional government spy) Christopher Marlowe fashioned in *Doctor Faustus* a frightening caricature of the aspiring and independent natural philosopher. Faustus sought secular power through philosophy:

> O, what a world of profit and delight,
> Of power, of honour, of omnipotence,
> Is promis'd to the studious artisan![3]

Behind these words lay Marlowe's anxieties about the political ramifications of the unlicensed pursuit of knowledge. John Donne's famous remark about the new philosophy was no mere reaction against mathematical or mechanistic descriptions of natural things, but an attack on the political morality of many now professing natural knowledge, and a defence of the primacy of the 'natural' social order:

> 'Tis all in pieces, all cohaerence gone;
> All just supply, and all Relation:
> Prince, Subject, Father, Sonne, are things forgot,
> For every man thinkes he hath got
> To be a Phoenix, and that there can be
> None of that kinde, of which he is, but hee.[4]

Francis Bacon transposed what he knew best and believed in most deeply into his form of natural philosophy – into his ingenious attempt to harness firmly any voluntaries in natural knowledge and to secure the monarchical civil order from the corrosive, even 'popular', political consequences of unmediated knowledge and unofficial knowledge-makers. He translated Tudor strategies of bureaucratic state management, late-Elizabethan political anxieties and social prejudices, and a principal intellectual resource of the landed gentry – the science of the common law – into his 'novum organum' for the paternal governance of the 'province' of knowledge.

When Bacon fell from political office in the early 1620s, so too did his slim chance of realising his grand political programme. Yet within years of his death his name and writings *as a natural philosopher* were being appropriated by men with interests and allegiances very different than his own.[5] There is no mystery in this. The political ambitions that had underpinned his vision of a natural philosophy were largely unknown, having been expressed in private letters and manuscript jottings not printed until much later. On the other hand, Bacon's published writings on the reform of natural philosophy had been purposefully designed to attract as many persons as possible to labour in his 'Great Reform' and, given that readers are always at liberty to read selectively, it is not surprising that readers would construe his words into different meanings. Further, it is always useful to claim an illustrious pedigree when advancing a novel and controversial programme, as were the Puritan social reformers of the 1640s and 1650s, and Robert Boyle, Thomas Sprat and their colleagues in the early years of the Royal Society of London.[6] It is more than reasonable to speculate that the man who described himself as 'a perfect and peremptory royalist'[7] would have been appalled by those Englishmen of the 1640s and 1650s who attacked the centralised institutions of the state and of learning while claiming to be faithfully following his natural philosophy. And despite the public rhetoric of the private gentlemen who came to dominate the early Royal Society of London (a society 'royal' in name only), it cannot be characterised as a 'Baconian' institution.[8] While inclined to minimise the significance of Bacon for the subsequent history of natural philosophy, most historians of the sciences feel comfortable in assigning great place to Robert Boyle and his 'experimental philosophy', stressing in particular the importance of his methodological protocols for natural knowledge won through experiment, which he elaborated and fiercely defended against critics. Yet Boyle's experimental practices (his 'technologies', as they recently have been described), deploy language, assumptions and structures of the common law: 'trials', 'credible witnesses', establishing 'matters of fact' and the like.[9] Bacon was the first to adapt the technical language and structures of the common law to a method for natural philosophy; Boyle, it seems, was the next to do so. Boyle was no lawyer, and although some familiarity with the workings of the common law was a prudent attainment for a gentleman with lands and much property, his redeployment of legal 'technologies' on this scale suggests more than a layman's familiarity. That Boyle 'took' these from Bacon's natural philosophy would be a rash assertion – independent discovery does often happen – but the question of where he learned them remains an intriguing one.

As for an institutional structure for the 'royal work' of pursuing natural knowledge, a highly influential and much-copied successor to Bacon's own proposals was formed elsewhere than in England. It is ironic, yet perhaps not surprising, that Bacon was highly regarded in Louis XIV's court and by Louis's faithful chief minister Colbert, who created for *his* imperial master an Académie des sciences 'in the manner suggested by Verulam'.[10] Francis Bacon's natural philosophy, after all, was a natural philosophy appropriate to a centralising and imperial state.

Appendix. A table of comparisons

The Works of Francis Bacon, edited by J. Spedding, R.L. Ellis and D.D. Heath in 7 volumes (London, 1857–61), was published in the United States of America in 15 volumes by Houghton, Mifflin and Co. (n.d.), Brown and Taggard (1860–4), and by Hurd and Houghton (1869–72), all three of which used the same press layout. The differences in volume and page references between the London and American imprints are so considerable that a comparison of them is required.

	UK edition Vol.		*US imprints* Vol.	
Life of ... Bacon, by Rawley	I	page 1	I	p. 33
General Preface ... by Ellis		p. 21		p. 61
Novum Organum Ellis' Preface		p. 71		p. 131
Instauratio magna		p. 119		p. 195
Praefatio		p. 125		p. 199
Distributio operis		p. 134		p. 212
Novum Organum		p. 149		p. 231
Praefatio		p. 151		p. 233
Liber primus		p. 157		p. 241
Liber secundus		p. 227		p. 341
Parasceve Preface		p. 369	II	p. 9
Descriptio historiae		p. 393		p. 43
Aphorismi		p. 395		p. 47
Catalogus		p. 405		p. 61
De augmentis scientiarum Preface		p. 415		p. 73
Partitiones scientiarum		p. 425		p. 87
Liber primus		p. 431		p. 97
Liber secundus		p. 485		p. 174
Liber tertius		p. 539		p. 251

A table of comparisons

	UK edition Vol.		*US imprints* Vol.	
Liber quartus	I	p. 579	II	p. 309
Liber quintus		p. 614		p. 358
Liber sextus		p. 651		p. 409
Liber septimus		p. 713	III	p. 7
Liber octavus		p. 745		p. 52
Liber nonus		p. 829		p. 174
Desiderata		p. 838		p. 187
On Writing in Cipher		p. 841	II	p. 499
Historia ventorum Preface	II	p. 3	III	p. 193
De historia ... monitum		p. 13		p. 205
Norma historiae praesentis		p. 17		p. 211
Historia ventorum		p. 19		p. 215
Aditus		p. 79		p. 299
Fragmentum ... abecedarium		p. 85		p. 306
Historia vitae et mortis Preface		p. 91		p. 315
Historia vitae et mortis		p. 101		p. 327
Historia densi et rari Preface		p. 229	IV	p. 9
Historia densi et rari		p. 241		p. 27
Inquisitio de magnete		p. 307		p. 119
Topica ... De luce et lumine		p. 313		p. 127
Sylva Sylvarum Preface		p. 325		p. 143
Sylva Sylvarum		p. 339		p. 159
(continued from 'Century VIII' in Vol. V p. 7)				
Table of the Experiments		p. 673	IV	p. 479
(repeated in Vol. V p. 165)				
Scala intellectus		p. 687	V	p. 177
Prodromi		p. 690		p. 182
Preface to Part II	III	p. 3		p. 187
Cogitationes de natura rerum		p. 11		p. 197
Preface De fluxu et refluxu		p. 39		p. 235
De fluxu et refluxu		p. 47		p. 247
Preface De principiis		p. 65		p. 271
De principiis		p. 79		p. 289
New Atlantis		p. 119		p. 347
Magnalia naturae		p. 167		p. 415
Preface to Part III		p. 171		p. 419
Cogitationes de scientia humana		p. 177		p. 427
Preface Valerius Terminus		p. 199	VI	p. 9
Valerius Terminus		p. 215		p. 25
Advancement of Learning I		p. 253		p. 77

Appendix

	UK edition Vol.		*US imprints* Vol.	
Advancement of Learning II	III	p. 321	VI	p. 171
Filum labyrinthi		p. 493		p. 413
De interpretatione naturae prooemium		p. 505		p. 429
Temporis partus masculus		p. 521	VII	p. 7
Delineatio et argumentum		p. 541		p. 33
Redargutio philosopharium		p. 557		p. 55
Cogitata et visa		p. 587		p. 97
Inquisitio legitima de motu		p. 621		p. 145
Calor et frigus		p. 641		p. 171
Historia soni et auditus		p. 653		p. 187
Phaenomena universi		p. 681		p. 223
Preface Descriptio globi		p. 715		p. 267
Descriptio globi intellectualis		p. 727		p. 285
Thema coeli		p. 769		p. 344
Sententiae XII		p. 780		p. 361
Aphorismi et consilia		p. 789		p. 371
Physiological and Medical Remains		p. 795		p. 379
The Great Instauration Prooemium	IV	p. 5	VIII	p. 15
Epistle Dedicatory		p. 9		p. 21
Preface		p. 13		p. 25
Plan of the Work		p. 22		p. 38
The New Organon Preface		p. 39		p. 59
Aphorisms Book I		p. 47		p. 67
Aphorisms Book II		p. 119		p. 167
Preparative		p. 249		p. 351
Catalogue		p. 265		p. 373
Dignity and Advancement Arguments		p. 275		p. 385
Book II		p. 283		p. 395
Book III		p. 336		p. 470
Book IV		p. 372	IX	p. 13
Book V		p. 405		p. 60
Book VI		p. 438		p. 107
Book VII	V	p. 3		p. 191
Book VIII		p. 31		p. 231
Book IX		p. 111		p. 345
History of the Winds		p. 137		p. 379
Preface History of Heavy and Light		p. 202		p. 468
Preface History of Sympathy & Antipathy		p. 203		p. 470
Preface History of Sulphur, Mercury & Salt		p. 205		p. 472
Fragment of Abecedarium naturae		p. 208		p. 475

A table of comparisons

	UK edition Vol.		US imprints Vol.	
History of Life and Death	V	p. 213	X	p. 7
History of Dense and Rare		p. 337		p. 177
Inquiry respecting the Magnet		p. 401		p. 267
Light and Luminous Matter		p. 407		p. 273
Thoughts on the Nature of Things		p. 417		p. 285
On the Ebb and Flow of the Sea		p. 441		p. 317
On Principles and Origins		p. 459		p. 341
Description of the Intellectual Globe		p. 501		p. 401
Theory of the Heaven		p. 545		p. 461
Index (to Volumes I–V)		p. 561		p. 481
History of ... King Henry VII	VI	p. 1	XI	p. 11
History of ... King Henry VIII		p. 265		p. 391
History of Great Britain		p. 271		p. 399
In felicem memoriam Elizabethae		p. 281		p. 411
Translation of *In felicem ...*		p. 305		p. 443
In Henricum Principem Walliae Elogium		p. 319	XII	p. 9
Translation of *Elogium*		p. 327		p. 19
Imago Civilis Julii Caesaris		p. 335		p. 27
Translation of *Julii Caearis*		p. 341		p. 35
Imago Civilis Augusti Caesaris		p. 339		p. 33
Translation of *Augusti Caesaris*		p. 347		p. 43
Additions ... to Camden's Annales		p. 349		p. 45
Essays (1625)		p. 365		p. 67
Fragment of an Essay on Fame		p. 519		p. 283
Essays (1597)		p. 521		p. 287
Manuscript collection		p. 535		p. 305
Essays (1612)		p. 537		p. 309
Attributed essays		p. 592		p. 383
De sapientia veterum		p. 605		p. 403
(continued at '*Fama*' in Vol. XIII p. 9)				
Translation of *De sapientia ...*		p. 687	XIII	p. 67
Touching a Holy War	VII	p. 1		p. 173
Of the True Greatness of Britain		p. 37		p. 219
Colours of Good and Evil		p. 65		p. 257
To Sir Henry Savill		p. 93		p. 291
Short Notes for Civil Conversation		p. 105		p. 305
Apophthegms		p. 111		p. 311
Promus of Formularies and Elegancies		p. 187	XIV	p. 7
Confession of Faith		p. 215		p. 41
Meditationes sacrae		p. 227		p. 59

Appendix

	UK edition Vol.		US imprints Vol.	
Translation of *Meditationes* ...	VII	p. 243	XIV	p. 81
Prayers		p. 255		p. 97
Translations of Certain Psalms		p. 263		p. 107
Christian Paradoxes		p. 287		p. 137
General Preface Professional Works		p. 301		p. 155
Maxims of the Law		p. 307		p. 163
Reading on the Statute of Uses		p. 389		p. 277
Use of the Law		p. 451		p. 361
The Commission of Bridewell		p. 505	XV	p. 7
Arguments of Law		p. 517		p. 23
Case of Impeachment of Waste		p. 527		p. 35
Lowe's Case of Tenures		p. 546		p. 62
Case of Revocation of Uses		p. 557		p. 78
The Council of the Marches		p. 567		p. 93
Chudleigh's Case		p. 613		p. 155
The Post-Nati of Scotland		p. 637		p. 189
De non procedendo Rege inconsulto		p. 681		p. 249
Preparation for the Union of Laws		p. 727		p. 313
The Office of Constables		p. 745		p. 335
Ordinances in Chancery		p. 755		p. 347
Appendix		p. 775		p. 373
Index (to Volumes VI–VII)		p. 783		p. 383

Notes

Introduction

1 William Rawley, 'The Life of the Honourable Author', *Works*, I, p. 11.

2 For example, see F.H. Anderson, *The Philosophy of Francis Bacon* (1948), and *Francis Bacon: His Career and His Thought* (1962); B. Farrington, *Francis Bacon: Philosopher of Industrial Science* (1949) and *The Philosophy of Francis Bacon* (1964); J.G. Crowther, *Francis Bacon: The First Statesman of Science* (1960); P. Rossi, *Francis Bacon: From Magic to Science* (1968); B. Vickers (ed.), *Essential Articles for the Study of Francis Bacon* (1968); M. Hesse, 'Francis Bacon', in C.C. Gillispie (ed.), *Dictionary of Scientific Biography*, I (1970), pp. 372–7; L. Jardine, *Francis Bacon: Discovery and the Art of Discourse* (1974); J. Lossee, *A Historical Introduction to the Philosophy of Science* (1980); G. Rees, *Francis Bacon's Natural Philosophy* (1984); D. Oldroyd, *The Arch of Knowledge: An Introductory Study of the History of Philosophy and Methodology of Science* (1986); C. Whitney, *Francis Bacon and Modernity* (1986); P. Urbach, *Francis Bacon's Philosophy of Science* (1987); and A. Perez-Ramos, *Francis Bacon's Idea of Science and the Maker's Knowledge Tradition* (1988). T.K. Rabb, 'Francis Bacon and the Reform of Society', in T.K. Rabb and J.E. Seigel (eds.), *Action and Conviction in Early Modern Europe* (1969), pp. 169–93, includes a useful survey of changing modern interpretations of Bacon.

3 For example, J.W. Allen, *English Political Thought, 1603–1660*, I (1938); J. Marwil, *The Trials of Counsel: Francis Bacon in 1621* (1976); J.J. Epstein, *Francis Bacon: A Political Biography* (1977).

4 See K. Wallace, *Francis Bacon on Communication and Rhetoric* (1943); L.C. Knights, 'Bacon and the Seventeenth-Century Dissociation of Sensibility', in his *Explorations* (1946); G. Williamson, *The Senecan Amble: A Study in Prose Form from Bacon to Collier* (1951); D. Bush, *English Literature in the Earlier Seventeenth Century* (1962); B. Vickers, *Francis Bacon and Renaissance Prose* (1968); L. Jardine, *Discovery and the Art of Discourse*; J. Pitcher (ed.), *The Essays* (1985); and J. Briggs, *Francis Bacon and the Rhetoric of Nature* (1989).

5 There are provocative hints towards this in W.S. Holdsworth, *A History of English Law*, V (1924), p. 239; P.H. Kocher, 'Francis Bacon on the Science of Jurisprudence', *Journal of the History of Ideas*, 18 (1957), pp. 3–26: reprinted in B. Vickers (ed.), *Essential Articles*; and in R.T. Simonds, 'Bacon's Legal Learning: Its Influence on His Philosophical Ideas', in I.D. McFarlane (ed.), *Acta Conventus Neo-Latini Sanctandreani: Proceedings of the Fifth International Congress of Neo-Latin Studies* (1986), pp. 493–501.

6 A tradition stretching from T.B. Macaulay's attack on Bacon's political morals in *The Edinburgh Review*, 132 (July 1837), pp. 1–104 and the ambivalence of the great mathematician Robert Leslie Ellis in his 'general preface' to the *Works*; see R.W. Church, *Bacon* (1884), and J.G. Crowther, *Statesman of Science* (1960).

7 S.R. Gardiner, 'Francis Bacon', *Dictionary of National Biography* (1885); J. Marwil, *Trials of Counsel*, esp. pp. 125, 134. Cf. J.J. Epstein, *Political Biography*, Chapter 1.

8 See the 'Puritanism and Science' and the 'Origins of the Royal Society' debates in the 1960s and early 1970s in the journal *Past and Present*; P. Wood, 'Francis Bacon and the "Experimentall Philosophy": A Study in Seventeenth-Century Methodology' (unpublished London M.Phil. thesis, 1978); and M. Hunter, *Science and Society in Restoration England* (1981) and *The Royal Society and its Fellows 1660–1700* (1982).

9 The fundamental importance of these issues is illustrated in S. Shapin and S. Schaffer, *Leviathan and the Air-Pump: Hobbes, Boyle and the Experimental Life* (1985), L.W.B. Brockliss, *French Higher Education in the Seventeenth and Eighteenth Centuries: A Cultural History* (1987) and J. Martin, 'Natural Philosophy and its Public Concerns', in S. Pumfrey, P. Rossi and M. Slawinski (eds.), *Science, Culture and Popular Belief in Renaissance Europe* (1991), pp. 100–18.

1. A statesman's responsibility

1 See J.K. McConica, *English Humanists and Reformation Politics* (1965), esp. Chapters 3 and 4.

2 See A. Grafton and L. Jardine, *From Humanism to the Humanities* (1986).

3 A.G. Dickens, *The English Reformation* (1967), p. 102; and see F. Caspari, *Humanism and the Social Order* (1954), Chapters 1–3.

4 On the authority of royal proclamations, see R.W. Heinze, *The Proclamations of the Tudor Kings* (1976), pp. 30–54; on the statute of proclamations itself, see pp. 153–77. See also G.R. Elton, 'Henry VIII's Act of Proclamations' [1960], in his *Studies in Tudor and Stuart Politics and Government*, I (1974), pp. 339–54.

5 See G.R. Elton, *Reform and Renewal: Thomas Cromwell and the Commonweal* (1973).

6 F. Caspari, *Humanism* and A.B. Ferguson, *The Articulate Citizen and the English Renaissance* (1965).

7 W.R.D. Jones, *The Tudor Commonwealth, 1529–1559* (1970) and W.K. Jordan (ed.), *The Chronicle and Political Papers of King Edward VI* (1966).

8 J.K. McConica, *Reformation Politics*, claimed that everyone was an 'Erasmian', really; see his Chapter 6: 'Official Erasmianism: The Work of Thomas Cromwell, 1535 to 1540'.

9 See G.R. Elton, *Reform and Renewal* (1973) and 'Reform and the "Commonwealth-Men" of Edward VI's Reign' (1979), in *Studies*, III (1983), pp. 234–53.

10 G.R. Elton, 'Thomas Cromwell Redivivus' (1977), in *Studies*, III, pp. 373–90; and see Q. Skinner, *The Foundations of Modern Political Thought, I: The Renaissance* (1978), esp. Chapter 6.

11 G.R. Elton, 'Thomas Cromwell Redivivus', p. 378.

12 See A.G. Dickens, *English Reformation*, pp. 122–5, 158, 168; 'If we seek for a fundamental book on Church and state which helped to inspire Cromwell's revolutionary ideas, we are more likely to find this in the *Defensor pacis* of Marsiglio' (p. 158). See also G.R. Elton, *Reform and Renewal* and 'The Political Creed of Thomas Cromwell' (1956), in *Studies*, II (1974), pp. 215–35.

13 Q. Skinner, *Foundations*, I, p. 65; P.O. Kristeller, *Iter Italicum: A Finding List of Uncatalogued or Incompletely Catalogued Humanistic Manuscripts of the Renaissance in Italian and other Libraries* (1967).

14 G.R. Elton, 'Political Creed', p. 228. Cromwell gave money for its publication to William Marshall, one of his propagandists, who translated *Defensor pacis*. Marshall's was the only edition (translated or not) printed in England before 1700; see the *Short Title Catalogues*.

15 F. Gilbert, 'Bernardo Rucellai and the Orti Oricellari: A Study on the Origin of Modern Political Thought', *Journal of the Warburg and Courtauld Institutes*, 12 (1949), pp. 101–31.

16 F. Gilbert, 'The Venetian Constitution in Florentine Political Thought', in N. Rubinstein (ed.), *Florentine Studies* (1968), pp. 463–500; H.C. Butters, *Governors and Government in Early Sixteenth-Century Florence, 1502–1519* (1985); and R. Finlay, *Politics in Renaissance Venice* (1980).

17 This story originated with Reginald Pole, one of Cromwell's most bitter enemies; see G.R. Elton, 'Political Creed', pp. 217–21.

18 See A. Gewirth (ed. and trans.), *Marsilius of Padua: 'The Defender of the Peace'*, 2 vols. (1951–6).

19 By these means, Marsiglio delimited the role of the ruler and defined his specific responsibilities as well as repudiating all claims for legitimate political sovereignty based on supra-mundane arguments; see M. Wilks, *The Problem of Sovereignty in the Later Middle Ages* (1963), esp. pp. 84–117; on Marsiglio's beliefs about law and on the rule of law, see A. Gewirth (ed.), *Marsilius*, I, pp. 132–66, 226–59.

20 *Discourse One*, I, i.

21 See S.T. Bindoff, 'The Greatness of Antwerp', in G.R. Elton (ed.), *The New Cambridge Modern History, II: The Reformation* (1958), pp. 50–69. The importance of the cloth trade and Antwerp to England's economy is

surveyed by D.M. Palliser, *The Age of Elizabeth: England under the Later Tudors, 1547–1603* (1983), pp. 237–65, 281–7.

22 24 Henry VIII, c. 12; my text is G.R. Elton, *The Tudor Constitution*, pp. 344–9.

23 G.R. Elton, *Reform and Renewal*, p. 68.

24 G.R. Elton, *ibid.*, Chapter 3: 'Cromwell and His Men'. Cromwell 'encouraged the assembling of a group of advisors whose chief usefulness lay in their ability to write propaganda and develop plans of reform' (p. 26); prominent among these men were Thomas Starkey, William Marshall and Richard Morison, all of whom recently had lived in northern Italy. See Elton's 'Reform by Statute: Thomas Starkey's *Dialogue* and Thomas Cromwell's Policy' (1968), in *Studies*, II, pp. 236–58; and W.G. Zeeveld, *Foundations of Tudor Polity* (1948), esp. Chapters 2 and 3.

25 G.R. Elton, *Reform and Renewal*, Chapter 4: 'The Instrument'; 'The reformers wanted new laws, and in England this could only mean new parliamentary statutes' (p. 66).

26 In the terms of *Defensor pacis*, these sorts of bills were practical necessities of a drive to create 'tranquillity' in the state.

27 *Reform and Renewal*, pp. 94–7. Elton identified ten 'commonweal' acts during 1532; nine more for 1533; twelve (and nine failed bills) in 1534; five acts (and twelve failed bills) in 1536; and in 1540, twenty-six acts (and fifteen failed bills).

28 G.R. Elton, *Reform and Renewal*, Chapter 6: 'Law Reform'.

29 See 23 Henry VIII, cc. 1, 11; 25 Henry VIII, c. 13; 27 Henry VIII, cc. 16, 19; 28 Henry VIII, c. 1; 32 Henry VIII, c. 12.

30 See 23 Henry VIII, cc. 2, 3, 6, 9, 13, 14, 15; 24 Henry VIII, cc. 5, 8; 25 Henry VIII, cc. 4, 6, 11; 27 Henry VIII, cc. 5, 11, 26; 28 Henry VIII, c. 9; 31 Henry VIII, cc. 1, 3, 6; 32 Henry VIII, cc. 4, 9, 20, 28, 30–4, 37.

31 G.R. Elton, *Reform and Renewal*, p. 143. See S.F.C. Milsom, *Historical Foundations of the Common Law* (1981), pp. 218–30; J.H. Baker, *An Introduction to English Legal History* (1979), pp. 230–57, *passim*; and A.W.B. Simpson, *A History of the Land Law* (1986), pp. 184–99.

32 Elton, *Tudor Revolution*, pp. 203–23; and *Tudor Constitution*, pp. 131–3, 136–43; and W.C. Richardson, *The History of the Court of Augmentations, 1536–1554* (1961), Chapters 3 and 4.

33 See W.C. Richardson, *Court of Augmentations* and *Tudor Chamber Administration, 1485–1547* (1952), pp. 305–449; and H.E. Bell, *An Introduction to the History and Records of the Court of Wards and Liveries* (1953), esp. pp. 16–132.

34 See W. MacCaffrey, *The Shaping of the Elizabethan Regime* (1968), pp. 27–40; and N.L. Jones, 'Elizabeth's First Year', in C. Haigh (ed.), *The Reign of Elizabeth* (1984), pp. 27–53. On the previous period, see J. Loach and R. Tittler (eds.), *The Mid-Tudor Polity, c. 1540–1560* (1980).

35 SP.12/1/7 and 8. The careers and activities of these men impinge so greatly upon each other, at least from the 1550s, that it would be very difficult to uncover who was 'influencing' whom, but it is clear they had much in

common and had remained allies during the Edwardian and Marian regimes.

36 M. Dewar (ed.), *A Discourse of the Commonweal of This Realm of England* (1969), esp. pp. 63–9, and 87–92.

37 Smith was not interested in condemning the morals of the rich, the powerful or the poor.

38 'Beyond the sea, they reward and cherish every man that brings in any new art or mystery whereby the people might be set to work with such thing as should both find [livings for] their workmen and also bring some treasure or other commodity into the country', in M. Dewar (ed.), *Discourse*, p. 88.

39 M. Dewar (ed.), *Discourse*, p. 91.

40 M. Dewar, *Sir Thomas Smith: A Tudor Intellectual in Office* (1964), pp. 56–78.

41 M. Dewar (ed.), *Discourse*, p. xxiv: *Calendar of State Papers, Domestic, 1547–80*, p. 20: Smith to William Cecil. Cecil possessed a copy of Smith's *Discourse*: *Calendar of Salisbury Manuscripts*, I, no. 255.

42 See C. Read, *Mr Secretary Cecil and Queen Elizabeth* (1955) and *Lord Burghley and Queen Elizabeth* (1960); R. Tittler, *Nicholas Bacon: The Making of a Tudor Statesman* (1976) and A. Simpson, *The Wealth of the Gentry, 1540–1600* (1961), Chapter 2: 'The Rising Lawyer: Sir Nicholas Bacon' (pp. 22–114). The biographies in Bindoff's and Hasler's volumes of *The House of Commons* are excellent guides to their careers in the Commons.

43 W. MacCaffrey's phrase, in *The Shaping of the Elizabethan Regime* (1968), p. 15.

44 On the critical problems facing the new government see W. MacCaffrey, *Shaping*, and N.L. Jones, 'Elizabeth's First Year'.

45 See P.L. Hughes and J.F. Larkin (eds.), *Tudor Royal Proclamations, II: 1553–1587* (1969), nos. 464 (1559); 493, 494, 496 (1562); 542 (1566); *et passim.*

46 See 5 Elizabeth c. 22; 8 Elizabeth cc. 3, 6, 15.

47 See 1 Elizabeth c. 14; 5 Elizabeth c. 2; 8 Elizabeth c. 13. Cecil's draft for a 1563 speech on 'fish days' and the like is SP.12/27/71. See also P.L. Hughes and J.F. Larkin (eds.), *Tudor Royal Proclamations*, II, no. 463 (1559), which prohibited the sale of ships to foreigners. The beginning of Lent yearly produced the heavy secular instrument of a royal proclamation forbidding eating of meat; see Hughes and Larkin, *passim*; and F.A. Youngs, jr., *Tudor Queens*, pp. 123–5 on 'political Lent'.

48 See J. Thirsk, *Economic Policy and Projects: The Development of a Consumer Society in Early Modern England* (1978), for an excellent discussion of the Tudor patent system and the Elizabethan and Jacobean government policies regarding them. Thirsk concluded of Cecil: 'His method of informing himself on foreign developments in industry and agriculture, his retention of advisers and fact-finders who combed England and Europe and drew up reports on the feasibility of domestic projects, all

bear strong traces of a method *derived from the years of Thomas Cromwell's administration.* Cecil pushed the method further, but the fundamental approach was the same' (p. 33); my emphasis. Cecil's papers in the British Library Lansdowne manuscript collection are full of such reports and petitions about 'projects'.

49 R. Tittler, *Nicholas Bacon*, pp. 109–10 and p. 225, nn. 31 and 32. Bacon's paper was entitled 'A Remembraunce howe to proceede to the understandinge of the state of her Majesties revenewe and of the reformacion of the unnecessarie expences thereof'.

50 R. Tittler, *ibid.*, pp. 112 and 226, nn. 36 and 38. On the coinage, see J.D. Gould, *The Great Debasement: Currency and the Economy in Mid-Tudor England* (1970); F.A. Youngs, jr., *Tudor Queens*, pp. 104–10; and C.E. Challis, *The Tudor Coinage* (1978). The Privy Council issued several proclamations about coinage in the autumn of 1560 and during 1561; P.L. Hughes and J.F. Larkin (eds.), *Royal Proclamations*, nos. 471, 472, 473, 475, 478, 480, 487, 488, 491, 492.

51 T.E. Hartley (ed.), *Proceedings in the Parliaments of Elizabeth I. Volume One: 1559–1581* (1981), p. 34.

52 *Ibid.*, pp. 49–50. 'Is it not (trowe you) a monsterous disguising to have a justicer a maynteyner, to have hym that should by his othe and dutie set forthe iustice and righte, against his othe and dutie offer iniurye and wronge, to have him that is specially chosen amongste a number by the prince to appease all brablings and controversies, to be a sower and maynteyner of strife and sedition amongst them, seeking his reputacion and opynion by leading and sweyinge of juryes accordinge to his will, acquiteinge some for gayne, inditeinge other for mallice, bearing with him as his servante, overthrowinge the other as his enemye, procuringe all questmongers to be of his liverye or otherwise in his daunger, that his winkes, frowninges and contenances may direct all inquestes?' (p. 50). J.S. Cockburn, *A History of the English Assizes, 1558–1714* (1971), gives a similar picture, pp. 153–261.

53 See R. Tittler, *Nicholas Bacon*, pp. 75–80, and G. Sanders (ed.), *Orders of the High Court of Chancery*, I, i (1845), pp. 20–56.

54 R. Tittler, *Nicholas Bacon*, pp. 107–8, quoting Huntington MS Ellesmere 2579, fos. 13v–14. Bacon repeated parts of this speech *verbatim* on subsequent occasions; see T.E. Hartley, *Proceedings*, pp. 190, 192–3 (1571) and pp. 464–5 (1576).

55 On the justices of the peace, see J.S. Cockburn, *Assizes*; J.H. Gleason, *The Justices of the Peace in England, 1558–1640* (1969), pp. 47–67, 96–115; and M.G. Davies, *The Enforcement of English Apprenticeship: A Study in Applied Mercantilism, 1563–1642* (1956), esp. pp. 161–228.

56 R. Tittler, *Nicholas Bacon*, p. 103, and S.T. Bindoff, 'The Making of the Statute of Artificers', *Elizabethan Government and Society* (1961), eds. S.T. Bindoff, J. Hurstfield and C.H. Williams, pp. 87, 90.

57 S.T. Bindoff, 'Statute of Artificers', p. 87. Bacon planned a complete overhaul of the existing statutes to weed out duplications, and to eliminate

superceded or redundant statutes. He submitted his plan to the Privy Council in 1575, proposing several committees each of four senior lawyers (whom he named) in order to carry out this bold programme; see SP.12/105/93.

58 5 Elizabeth cc. 7, 12, 24, 25, 26, 27. Three more law reform statutes were passed in the 1566 parliament: 8 Elizabeth cc. 2, 4, 5. On the assessment of bills as 'official' or 'private', see G.R. Elton, 'Enacting Clauses and Legislative Initiative, 1559–1581' (1980), *Studies*, III, pp. 142–55, and *The Parliament of England, 1559–1581* (1986), pp. 43–61.

59 T.E. Hartley, *Proceedings*, pp. 80–6 (1563), esp. p. 83; and also p. 192 (1571).

60 C. Read (ed.), *William Lambarde and Local Government* (1962), p. 76; my emphasis. Lambarde's various charges to the Kent juries of presentment contain striking declarations of the active role and the 'ministry' of the JPs: he regarded himself and his fellow JPs as physicians to the body politic, and royal justice as physic. *His* view of the office of JP was that of the central officers of State: 'it pleaseth her Highness to use us as the mouth of her laws in this behalf (for indeed we ought to be *lex loquens*, a speaking law)' (24 April 1582), see C. Read, p. 69. There are many indications that the clergy regarded the magistrates and themselves as being in harness together; see P. Collinson, *The Religion of Protestants: The Church in English Society, 1559–1625* (1982), pp. 141–88.

61 T.E. Hartley, *Proceedings*, pp. 184–5.

2. The young statesman

1 See William Rawley, *The Life of the Honourable Author* (1657), *Works*, I, pp. 3–4. The extremely slim evidence about the schooling given in Nicholas Bacon's household is reviewed in R. Tittler, *Nicholas Bacon: The Making of a Tudor Statesman* (1976), pp. 58–63. A few of Tittler's claims exceed the evidence: his belief that Thomas Blundeville, a mathematical populariser, had served as a tutor in Bacon's household rests on the unsubstantiated claim of E.G.R. Taylor, *The Mathematical Practitioners of Tudor and Stuart England* (1954), p. 173; Blundeville's patron was the Earl of Leicester. Nor can it be argued on the basis of the begging dedication of his son's *Panometria* (1571) that Leonard Digges' 'work on optics and mensuration was carried out with Bacon's patronage' (p. 58): it is more likely Digges surveyed some lands for the Lord Keeper.

2 Cambridge scholars who tutored undergraduates were expected to be, quite literally, *in loco parentis*. Bacon's elder sons had gone to Trinity College, as well, but long before Whitgift migrated from Peterhouse, *via* the mastership of Pembroke in 1567.

3 On the religious and scholarly strife in Cambridge in the 1560s and 1570s, see H.C. Porter, *Reformation and Reaction in Tudor Cambridge* (1958), esp. pp. 101–82. Whitgift was thirty-five and the Regius Professor of Divinity in 1567 when his preaching delighted the queen, who thereafter

encouraged his rapid promotions. Whitgift's long controversy with the Puritan Thomas Cartwright is discussed by H.C. Porter, *Tudor Cambridge*, and by P. Collinson, *The Elizabethan Puritan Movement* (1967), pp. 122–5; see also *The Works of John Whitgift DD*, 3 vols., ed. J. Ayre, (1851), much of which is devoted to Whitgift's lengthy refutations of Cartwright's equally lengthy polemics.

4 See P. Collinson, ' Sir Nicholas Bacon and the Elizabethan Via Media', *Historical Journal*, 23 (1980), pp. 255–73.

5 J. Whitgift, *Works*, III, p. 568.

6 *Ibid.*, *Works*, I, pp. 11–12.

7 On the Cambridge curriculum, see L. Jardine, *Francis Bacon: Discovery and the Art of Discourse* (1974), *passim*. Jardine makes extensive claims for the importance of Peter Ramus in Francis Bacon's education and outlook. Ramus was popular with Puritan scholars in Cambridge, especially after his martyrdom in the St Bartholemew's Day Massacre in 1572, and Francis Bacon (like many students in Cambridge) probably was exposed to some of his work on logic, attacking Aristotle. However, 'it would appear that the influence of the language and method of Peter Ramus on the Cambridge Puritans can be exaggerated' (H.C. Porter, *Tudor Cambridge*, p. 225n.), and see J.K. McConica, 'Humanism and Aristotle in Tudor Oxford', *English Historical Review*, 94 (1979), pp. 291–317. Whitgift was adamantly opposed to Ramus, detesting his doctrines about logic nearly as much as those of doctrinal non-conformists, and we can hardly expect his personal pupils to have been unaffected by his convictions. The undergraduate curriculum was based squarely upon Aristotle's works of logic, but we grievously underestimate the Elizabethan universities if we accept the calumny that the curriculum was dull, monolithic and antique. See the work of C.B. Schmitt; esp. *Studies in Renaissance Philosophy and Science* (1981), *Aristotle and the Renaissance* (1983) and *John Case and Aristotelianism in Renaissance England* (1983).

8 J. Foster, *The Register of Admissions to Gray's Inn, 1521–1589* (1889), p. 48.

9 On the organisation of the Inns and the education conducted there, see W.C. Richardson, *A History of the Inns of Court* (n.d. 1977?), pp. 15–166; W.R. Prest, *The Inns of Court under Elizabeth I and the Early Stuarts, 1590–1640* (1972) and J.H. Baker (ed.), *Spelman's Reports*, II (1978), pp. 123–37.

10 Robert Dudley tried to have Cecil disgraced and cast down from the Privy Council in the early 1560s. They learned subsequently to work together but their relationship was never cordial: see W.T. MacCaffrey, *The Shaping of the Elizabethan Regime* (1968), pp. 71–101. The politicking of the Privy Councillors is discussed by MacCaffrey, *Shaping*, *passim*, and in *Queen Elizabeth and the Making of Policy, 1572–1588* (1981), pp. 431–62; Simon Adams, 'Eliza Enthroned? The Court and its Politics', in C. Haigh (ed.), *Reign of Elizabeth* (1984), pp. 55–77; and M.B. Pulman, *The Elizabethan Privy Council in the Fifteen-seventies* (1971).

11 Leicester was especially good at marshalling literary support for 'Godly' causes. He recognised the usefulness of print, and he was willing to provide pensions and positions to men who could write persuasively. He aptly has been called the great 'patron of letters' in Elizabethan England: see E. Rosenberg, *Leicester: Patron of Letters* (1955). The tremendous respect and expectations he and his policies generated among the scholarly is signalled by his receipt of more dedications of books than anyone except the queen herself; see F.B. Williams (ed.), *An Index of Dedications and Commendary Verses in English Books before 1641* (1962), p. 58.

12 Sidney corresponded regularly with Catholic and Protestant scholars, politicians and poets in France, the Low Countries, the Rhineland, Geneva and Vienna. He was especially close to several of the Huguenot courtier-poets of Henri of Navarre, in particular Philippe Du Plessis Mornay, with whom he struck up a lasting friendship. See J.A. Van Dorsten, *Poets, Patrons, and Professors: Sir Philip Sidney, Daniel Rogers, and the Leiden Humanists* (1962), esp. pp. 27, 48–9, 57–8; J.M. Osborn, *Young Philip Sidney, 1572–1577* (1972) and R. Howell, jr., 'The Sidney Circle and the Protestant Cause in Elizabethan Foreign Policy', *Renaissance and Modern Studies*, 19 (1975), pp. 31–46.

13 A.C. Hamilton, *Sir Philip Sidney: A Study of his Life and Works* (1977), p. 1; and Van Dorsten, *Poets, Patrons, and Professors*, pp. 48–9.

14 Sir Nicholas Bacon advocated this sort of experience for young law students preparing for a career as royal servants. His recommendations as far back as 1540 to King Henry VIII about the establishment of a special royal Inn of Court are printed by W.C. Richardson, *A History of the Inns of Court*, pp. 409–32.

15 See Sir Amias Paulet, *Copy-Book of Letters*, ed. O. Ogle (1866); the letters cover the period from May 1577 until January 1578. On two occasions, Paulet assured the Lord Keeper that Francis, 'my companion', was in good health. Young Bacon was with Paulet in Poitiers when the French court met the diplomats and courtiers of Henri, king of Navarre (among whom were the poets Du Plessis Mornay and Sallust Du Bartas); see O. Ogle (ed.), *Copy-Book*, pp. 77–8, 129–30.

16 6 February 1576; *Calendar of State Papers, Foreign, Elizabeth*, p. 242.

17 On the *académies* in general, see F.A. Yates, *The French Academies of the Sixteenth Century* (1947).

18 F.A. Yates, *French Academies*, pp. 69–76, 198–235.

19 Daniel Rogers, quoted in Van Dorsten, *Poets, Patrons, and Professors*, p. 66. On Sidney's 'Areopagus', which included Fulke Greville, Sir Edward Dyer, Daniel Rogers, Gabriel Harvey and Edmund Spenser, see J.E. Phillips, 'Daniel Rogers: A Neo-Latin Link Between the Pleiade and Sidney's "Areopagus"', in J.E. Phillips and D.C. Allen, *Neo-Latin Poetry of the Sixteenth and Seventeenth Centuries* (1965).

20 Lady Anne became more extreme in the following years, and increasingly patronised and protected young radical clerics; eventually she attended the

'godless' court only to petition in favour of imprisoned preachers; see P. Collinson, *Puritan Movement*, p. 257.

21 On the general political scene from the English council's point of view, see W.T. MacCaffrey, *The Making of Policy*, pp. 267–301. On the 1581 parliament, see J.E. Neale, *Queen Elizabeth I and Her Parliaments 1559–1581* (1957) and now G.R. Elton, *The Parliament of England, 1559–1581* (1986).

22 The inner barristers studied 'by the space of seven years, or thereaboutes' until 'growing ripe in the knowledge of the lawes, and approved withall to be of honest conversation, they are either by the general consent of the Benchers, or Readers, being the most ancient, grave, and judiciall men of everie Inne of the Court, or by the speciall priviledge of the present reader there, selected and called to the degree of Utter Barresters', John Stow, *A Survay of London*, I (1598), p. 78.

23 James Spedding assumed Burghley's support for Bacon. During 1583, Bacon remained closely associated with the Leicester House and Puritan circles in London. Letters from Walsingham's secretary show that Francis often went to the sermons of Walter Travers, the famous Puritan lecturer at the Temple, in the company of his mother: *Cal. S.P. Foreign, Elizabeth*, 20 November 1583.

24 Sir Philip Sidney, *An Apology for Poetry, or, The Defense of Poesy*, ed. G. Shepherd (1967), p. 104: lines 2–8, 34–8.

25 See the article on Bacon in P. Hasler (ed.), *The House of Commons, 1558–1603*, I (1981), p. 375. The coolness of Burghley towards Bacon, and his apparent unwillingness in later years to provide him with a government sinecure may seem puzzling, yet given Bacon's stubborn adherence to Burghley's political rivals, the Lord Treasurer's suspicions seem sensible.

26 P. Collinson, *Puritan Movement*, p. 272. On this important period, see P. Collinson, *Puritan Movement*, pp. 243–90; P.M. Dawley, *John Whitgift and the Reformation* (1955), pp. 161–94; W.T. MacCaffrey, *Making of Policy*, pp. 97–118; and J.E. Neale, *Elizabeth I and Her Parliaments, 1584–1601* (1957), pp. 13–101.

27 Two others who wrote treatises were Robert Beale and Thomas Digges. Beale, who had travelled with Sidney to the Continent in 1572 and was now a clerk of the Privy Council, wrote a long tract quoted extensively by J.E. Neale, *Elizabeth and Her Parliaments, 1584–1601*, pp. 66–9. Digges, the mathematician, was a member of Leicester's household; he wrote his 'Discourse' in January 1585 (Neale, *Parliaments, 1584–1601*, pp. 44–5). Bacon's '*Letter*' has no date, but internal evidence suggests it was written before April of 1585, and probably within the life of this parliament. For the text of Bacon's 'Letter of Advice', see J. Spedding, *Letters*, I, pp. 47–56.

28 In *Letters*, I, pp. 43–6, Spedding discussed the several manuscript copies.

29 On the Throckmorton plot, see W.T. MacCaffrey, *Making of Policy*, pp. 302–47; and E. St John Brooks, *Christopher Hatton* (1946),

pp. 251–9. On the Parry plot, see J.E. Neale, *Parliaments, 1584–1601*, pp. 40–1, 48–9 and on Mildmay's speech, p. 28. Many Protestant gentlemen signed a 'Bond of Association' by which they would become vigilantes against Catholics if Elizabeth was murdered. On this bond, see D. Cressy, 'Binding the Nation: The Bonds of Association, 1584 and 1696', in D.J. Guth and J.W. McKenna (eds.), *Tudor Rule and Revolution* (1982), pp. 217–34.

30 Mildmay had declared: 'We should give them cause to think that we had wholly secluded them from our society, not accounting them as natural-born Englishmen, and thereby drive a desperation into them', J.E. Neale, *Parliaments, 1584–1601*, p. 57.

31 In February 1585; J.E. Neale, *Parliaments, 1584–1601*, pp. 65–6.

32 'Having long been smothered and mowed down by the differing Tyrannies of Spain and Rome, shall we not be confident they would ... both stir up those benumbed Soveraignties which onely bear the name of free Princes to effect their own Manumissions ... ?', Fulke Greville's *Life of Sidney*, quoted by R. Howell, jr., 'The Sidney Circle', p. 39.

33 *Letters*, I, p. 55. Other Puritan gentlemen beside Sidney were strong advocates of a similar naval strategy against Spain's colonies; see K.R. Andrews, *Trade, Plunder and Settlement: Maritime Enterprise and the Genesis of the British Empire, 1480–1630* (1984), *passim.*

34 Sallust Du Bartas wrote to Bacon in 1584 asking him to help in the printing of the second part of his *La Création du monde*, the first part of which Sidney had just translated. Du Bartas hoped as well that Bacon would convince the queen to favour his poetry. See Du Bartas, *Works*, ed. U.T. Holmes, et al., I (1938), Appendix B, p. 201.

35 See P. Collinson, *Puritan Movement*, pp. 291–316.

36 After Leicester's death (September 1588), Walsingham worked very closely with the conservative Sir Christopher Hatton on security matters; C. Read, *Mr Secretary Walsingham and the Policy of Queen Elizabeth*, III (1925), p. 336. Lady Anne Bacon, however, was increasingly involved with Puritan clerics. Not only was she the financial supporter of several leading radicals but she was also engaged in their massive compilation about the woeful state and competence of England's parish clergy, known as *The Second Part of a Register*, ed. A. Peel (1915), p. 13.

37 On the tortuous slide of England and Spain into war, and on the Leicester expedition into the Low Countries, see W.T. MacCaffrey, *Making of Policy*, pp. 302–401.

38 On Travers and his Middle Temple lecturership controversy with Richard Hooker, see S.J. Knox, *Walter Travers: Paragon of Elizabethan Puritanism* (1962), pp. 70–88.

39 The political conservatism of the civil lawyers is surveyed in B.P. Levack, *The Civil Lawyers in England, 1603–1642: A Political Study* (1973).

40 P. Hasler (ed.), 'Francis Bacon', in *House of Commons*, I, p. 375. After Bedford's death, the Earl of Warwick managed his patronage on behalf of the young heir.

41 Lady Anne Bacon was shocked, nonetheless, and hereafter was deeply suspicious of Francis' religious sincerity; see her comments in letters to her son Anthony: e.g., *Letters*, I, pp. 112–15.

42 William Camden, *Annals* (1635 edition), p. 394. The state papers and C. Read's biography amply illustrate the size and vigour of Walsingham's intelligence network. Among Walsingham's foreign 'intelligencers' was Anthony Bacon, whose surviving correspondence with Walsingham begins after his arrival in France and continues through the 1580s; see the 'Bacon Papers' in the Lambeth (Tenison) MSS. Walsingham's 'usual medium for dealing with his spies' (C. Read, *Walsingham*, II, p. 336) was one of his secretaries, Thomas Phelippes, a brilliant decipherer who, in 1586, had been recommended for advancement in the secretary's office by Francis Bacon, whose companion Phelippes had been in previous years; see SP.12/194/8 and 21. On the government's anxieties about Jesuits and priests, see C. Read, *Walsingham*, *passim*, and W.T. MacCaffrey, *Making of Policy*, esp. pp. 144–53.

43 Bacon probably was working for Walsingham and other Privy Councillors in their capacity as members of the Court for Ecclesiastical Causes (the 'High Commission'). See R.G. Usher, *The Rise and Fall of the High Commission* (1913), ed. P. Tyler (1968), and G.R. Elton, *Tudor Constitution*, pp. 221–32. 'The watching of recusants, the warding of priests, and the banishing of Jesuits and priests was turned over ... to the common-law members of the Commission' (R.G. Usher, *High Commission*, p. 88). This was not a regular ecclesiastical appellate court, but concerned itself with enforcing discipline and submission to the royal supremacy. It therefore was akin to the Court of Star Chamber in its general purpose and in being a manifestation of the judicial powers of Privy Councillors and senior government officers (who, in this case, included senior members of the queen's legal and judicial establishment generally).

44 SP.12/203/57 (September 1587). Attorney-General Popham and Solicitor-General Egerton were members of the High Commission, and his association with them suggests Bacon's legal advice was well regarded. The surviving letter was written by William Waad, Clerk to the Privy Council, formerly Walsingham's private secretary, and a young companion of Sir Amias Paulet (and Francis Bacon) in France in 1576.

45 R.J. Fletcher (ed.), *The Pension Book of Gray's Inn* (1901), I, pp. 72, 77. Bacon chose to lecture on Church advowsons; see B.L. Stowe MS 424, fos. 145–50.

46 *Acts of the Privy Council, 1588*, pp. 235–6, and see SP.12/215/19. The three High Commissioners were Hopton, the lieutenant of the Tower, Peter Osborne, Lord Treasurer's Remembrauncer of the Exchequer, and Thomas Wrothe, an alderman of London. Among the others named with Bacon were Richard Young (JP for Middlesex and Customer of London) and Richard Topcliffe, the infamous anti-Catholic zealot, informer and (sometime) official torturer of Jesuit prisoners.

47 *Acts of the Privy Council, 1588*, p. 417 (27 December).
48 P. Hasler (ed.), *House of Commons*, I, p. 189; and R. Somerville, *History of the Duchy of Lancaster*, I (1953).
49 See *Letters*, I, pp. 95–102 for the text.
50 A 'reversion' was the right to become (or nominate) the next occupant of an office; reversions were usually purchased, and a patent of reversion provided one with a negotiable asset. Bacon's reversion is recorded in the Signet Office Docquet book: P.R.O., S.O. 3/1, fo. 211v (November 1589). The bill was apparently procured by the Lord Treasurer, Burghley, but his initial promotion of the grant or its active support cannot be inferred from this internal clerical register.
51 C.L. Schofield, *A Study of the Court of Star Chamber*, (1900), pp. 62–3. More generally, see G.R. Elton, *Tudor Constitution*, pp. 163–87.
52 William Hudson (a Jacobean Clerk of Star Chamber), *A Treatise on the Court of Star Chamber*, III, Chapter 23; printed in F. Hargrave (ed.), *Collectanea Juridica*, II (1792), p. 219.
53 See W. Pierce, *Historical Introduction to the Martin Marprelate Tracts* (1908).
54 The Privy Councillors, in Star Chamber, were anxious enough in 1586 to have issued a sweeping decree tightly regulating the printing trade; they spoke of 'sundry intolerable offences, troubles and disturbances [that] have happened, as well in the Church as in the civil government of the state' (G.R. Elton, *Constitution*, pp. 182–7). Whitgift appointed Dr Richard Bancroft to oversee both the hunting of Marprelate and the propaganda counter-attack upon him. Bancroft's authoritarian zeal resulted in poisonous, hysterical and often obscene attacks not only against Marprelate, but Puritans and reformers in general; for example, see his Paul's Cross Sermon (printed March 1589) and Thomas Nashe's anti-Martin pamphlets of 1589–90. See also M. Maclure, *The Paul's Cross Sermons, 1534–1642* (1958), pp. 72–5.
55 My text is *Letters*, I, pp. 74–95. Bacon's 'Advertisement' did circulate. The anonymous author of *A petition directed to her most excellent maiestie* (Middleburg, 1591) argued that 'a learned man, and friend to the Bishops noteth as abuses, their urging of Subscription, their oath ex officio, their excommunication for trifles, and easy silencing of Ministers', and he printed a marginal reference to 'Advert[isement] To[uching] the Church of England', not printed' (p. 6). There are close similarities between the 'Advertisement' and Bishop Cooper's *An Admonition to the People of England* (1589), particularly touching the social dangers of unbeneficed preachers. In 1604, Bacon published *Certain Considerations Touching the Better Pacification and Edification of the Church*, in which the arguments of his 1589 tract are reiterated; *Letters*, III, pp. 103–27.
56 The imagery of disease and wounds was repeated several times: what was being wounded were the 'bonds of peace' surrounding the body politic; see Marsiglio, *Defensor pacis*.
57 'He who manages a peace without reviewing the arguments of the

dissidents, [unlike] he who unites by equity, disappoints the minds of men of the sweetness of peace'.

58 *Letters*, I, pp. 76–7.

59 'An Advertisement Touching Private Censure', in *The Alnwick MS*, ed. F. Burgoyne (1904), p. 32. The manuscript is undated but its content, the fact that it shares several phrases with the dated 'Advertisement Touching the Controversies' (which follows it directly in the MS book), and the fact that all the other pieces in this manuscript of *c.* 1600 are in accepted chronological order lead me to place it here.

60 'Advertisement Touching the Controversies', *Letters*, I, pp. 82–3. This conclusion exactly follows an argument of Whitgift's against Thomas Cartwright in 1574; see *The Works of John Whitgift*, I, pp. 5–6.

3. Business of state

1 *Letters*, III, p. 143 (1604). Robert Devereux awaits both a modern scholarly biographer and the close scrutiny of historians. Paul Hammer's forthcoming Cambridge Ph.D. thesis should be of great value.

2 And after the spring of 1587, the Puritan William Davison, a Secretary of State, lost office and all political power when Elizabeth blamed him for disobeying her and ordering the execution of Mary, Queen of Scots.

3 This view is amply confirmed by George Peele, in his *An Eglogue. Gratulatorie* (1589), which he dedicated to 'the right honourable and renowned *Shepherd of Albion's Arcadia*: Robert, Earl of Essex' – an unmistakable reference to Sidney's great prose romance. See also the dedicatory remarks of Richard Harvey, in *A theologicall discourse of the Lamb of God* (1590), and those of Sir Richard Dallington, in *Hypnerotomachia* (1592).

4 Allies were soon placed at court: Daniel Rodgers, Sidney's poetic and diplomatist friend, in May 1587 became a clerk of the Privy Council, and was quickly joined by Sir Thomas Smith, Essex's private secretary since 1584.

5 Hastings to Essex, 9 September 1588; Claire Cross (ed.), *Letters of Francis Hastings* (1969), pp. 39–40.

6 Henry Lake to Sir Henry Sidney, *Sidney Papers*, I, p. 232. 'There is little doubt that for a year or so [*c.* 1588–90] Essex went out of his way to please some of the most extreme puritan elements', P. Collinson, *Puritan Movement*, p. 444. Working from F.B. Williams, *Index of Dedications*, p. 53, I estimate that a high proportion of the works dedicated to Essex in these years were from Puritan authors.

7 E.g., William Gager and Alberico Gentili. The pattern of Gentili's dedications is instructive: in 1582 – to Leicester; in 1585 – Philip Sidney; 1587 – Walsingham; in 1589 and 1590 – Essex (who in the latter year stood as godfather to Gentili's son); see E. Rosenberg, *Leicester, passim.*

8 After 1586, a 'peace-party' centred on Burghley and included Hatton,

Crofts, Buckhurst and Cobham; see C. Read, *Walsingham*, III, p. 146. With the 'Portugal Voyage' of mid-summer 1589, in which forces led by Essex raided Lisbon, Essex's reputation as a warrior hero was confirmed: see R.B. Wernham, *After the Armada: Elizabethan England and the Struggle for Western Europe, 1588–1595* (1984).

9 10 December 1589, *Cal. Salisbury MSS*, III, p. 445.

10 See P. Williams, *The Council in the Marches of Wales under Elizabeth* (1958), pp. 280–9.

11 *Cal. Salisbury MSS*, IV, p. 74.

12 E.g., Francis Bacon to T. Phelips: SP.12/238/138.

13 SP.12/239/114.

14 See P. Clark (ed.), *The General Crisis of the 1590s* (1985), and D.M. Palliser, *The Age of Elizabeth*. On the other hand (and looking slightly ahead), the costs involved in subduing Ireland during the 1590s were to prove very considerable.

15 See P. Bowden, 'Statistical Appendix', in J. Thirsk (ed.), *The Agrarian History of England & Wales*, IV, pp. 814–55; and D.M. Palliser, *Age of Elizabeth*, pp. 386–7. For the disasters of the 1590s, see also P. Clark (ed.), *General Crisis*, esp. pp. 3–66.

16 J. Walter and K. Wrightson, 'Dearth and the Social Order in Early Modern England', *Past and Present*, 71 (1976), p. 28.

17 4 November 1586; J. Goring and J. Wake (eds.), *Northamptonshire Lieutenancy Papers, 1580–1614* (1975), p. 24. 'The governing elite was drawn together above all by fear of the crowd' (P. Slack, 'Books of Orders: The Making of English Social Policy, 1577–1631', *Trans. R.H.S.*, 5th ser., 30 (1979), p. 17). On the problems of keeping order in times of price inflation, see P. Slack, 'Book of Orders'; J. Walter and K. Wrightson, 'Dearth and the Social Order' and, more generally, K. Wrightson, *English Society, 1580–1680* (1980), Chapter 6: 'Order'.

18 On Tudor proclamations, see R.W. Heinze, *Proclamations of the Tudor Kings* (1976) and F.A. Youngs, *Proclamations of the Tudor Queens* (1976).

19 P.L. Hughes and J.F. Larkin, *Tudor Royal Proclamations*, III, p. 61 (no. 726).

20 *Ibid.*, nos. 715, 716, 736, 740, 745, 746. On 'vagabonds' in general, see P. Slack, 'Vagrants and Vagrancy in England, 1598–1664', *Economic History Review*, 2nd ser., 27 (1974), pp. 360–79, and A.L. Beier, *Masterless Men: The Vagrancy Problem in England, 1540–1640* (1985).

21 P. Slack, 'Poverty and Social Regulation in Elizabethan England', in C. Haigh (ed.), *The Reign of Elizabeth*, p. 224.

22 *A second sound, or warning of the trumpet unto judgement. Wherein is prooved, that all the tokens of the Latter day, are not onlie come, but wel neere finished, with an earnest exhortation, to be in continuall readinesse* (1589), p. 21. In 1590, Marten wrote a tract as part of the anti-Puritan campaign of Whitgift and Bancroft: *A reconciliation of all the pastors and*

cleargy of this church of England (1590). On apocalyptic beliefs, see R. Bauckham, *Tudor Apocalypse: Sixteenth Century Apocalypticism, Millenarianism and the English Reformation* (1978).

23 Marten, *A second sound*, p. 25.

24 On 'moral reformation', see K. Wrightson's unpublished Cambridge Ph.D thesis: 'The Puritan Reformation of Manners' (1973). 'There seems to have been widespread support for regulation only when the "vices" posed threats to the peaceful government and good ordering of the realm' (Joan Kent, 'Attitudes of Members of the House of Commons to the Regulation of "Personal Conduct" in Late Elizabethan and Early Stuart England', *Bulletin of the Institute for Historical Research*, 46 (1973), p. 62).

25 See J. Thirsk, *Economic Policy and Projects: The Development of a Consumer Society in Early Modern England* (1978), and E.W. Hulme, 'The History of the Patent System under the Prerogative and at Common Law', *Law Quarterly Review*, 12 and 16 (1896, 1900); both Thirsk and Hulme concentrate on 'industrial' patents, but see F.A. Youngs, *Tudor Queens*, pp. 136–55, and C. MacLeod, *Inventing the Industrial Revolution: The English Patent System, 1660–1800* (1988), pp. 1–19, for discussion of patents in general. 'The use of patents and licences in local administration was ... sapping [justices'] prestige and undermining their control of local affairs, particularly their control of local taxation' (A.H. Smith, *County and Court: Government and Politics in Norfolk, 1558–1603* (1974), p. 246).

26 'There appears to have been a marked coincidence in Norfolk between religious radicalism [i.e., Puritanism] and the opposition to patentees' (A.H. Smith, *County and Court*, p. 317).

27 See my 'Sir Hugh Plat: An Elizabethan Scientist' (unpublished M.A. thesis, University of Texas, 1982) for discussion of the 'Godly' convictions of Plat and his friends.

28 Leonard Mascall, *The first booke of cattell, wherein is shewed the government of Oxen, Kine, Calves ... with divers approved remedies* (1587); Robert Payne, *The vale mans table. Herein is taught ... How to drain moores* (1583).

29 L. Mascall, *A booke of fishing with hooke & line, and of all other instruments thereunto belonging. Another of sundrie engines and trappes to take polecats* (1590).

30 Sir Thomas Chaloner, *A Shorte Discourse of the most rare and excellent vertue of Nitre* (1584); John Banister, *A Compendious Chyrurgerie* (1585); Thomas Newton, *The olde mans dietarie* (1586); and S. Kellwaye, *A Defensative Against the Plague* (1593).

31 John Taverner, *Certaine experiments concerning fish and fruite* (1600).

32 John Partridge, *The widowes treasure, plentifully furnished with secretes in physicke. Hereunto are adjoyned sundry pretie practices in cookerie* (1585: reissued 1586, 1588, 1595, etc.).

33 Hugh Plat, *A discoverie of certaine English wants* (1595); and *Diverse new sorts of soyle* (1594).

34 Hugh Plat, *A briefe apologie of certaine new inventions* (1593).
35 Sir Thomas Chaloner, *Vertue of Nitre*, 'Printer to Reader'.
36 *The first booke of cattell*, 'Dedication'.
37 John Taverner, *Certaine experiments*, 'To the Reader'.
38 *A discoverie of certaine English wants*, sig. A3.
39 *Ibid.*, sig. A3v-A4.
40 P. Collinson, *Puritan Movement*, p. 443.
41 P. Collinson, *The Religion of Protestants* (1982), Chapter 6: 'Voluntary Religion: Its Forms And Tendencies', p. 268.
42 See C. Hill, 'The Ratsbane of Lecturing', in his *Society and Puritanism in Pre-Revolutionary England* (1964), pp. 70–123: 'By 1583 nearly 1 in 3 of London's parishes had lectureships' (p. 110).
43 P. Collinson, *Religion of Protestants*, p. 190.
44 'Christopher Hill must be right to locate puritanism in the ranks of the economically independent' (P. Collinson, 'The Godly' (1966), published in his *Godly People* (1983), p. 4).
45 P. Collinson, 'The Church and the New Religion', in C. Haigh (ed.), *Reign of Elizabeth*, p. 182. My emphasis.
46 'The rather ramshackle edifice of hermetic and alchemical writings was stimulating, even exhilarating, to generations breaking away from the highly organized didactic regimen of scholasticism. It is not accidental that alchemy was taken up by those searching for new modes of philosophical expression, such as ... *radical reformers within the church*', C. Webster, 'Alchemical and Paracelsian Medicine', in C. Webster (ed.), *Health, Medicine and Mortality in the Sixteenth Century* (1979), p. 314; my emphasis. A good example of a Puritan Paracelsian tract is R.B., *The Difference between the Ancient Physicke, first taught by the godly forefathers, consisting in peace and concord: and the latter Physicke proceeding from Idolaters, Ethnickes, and Heathen: as Gallen, and such other consisting in dualitie, discorde, and contrarietie. And wherein the naturall Philosophie of Aristotle doth differ from the trueth of God's worde, and is injurious to Christianitie and sounde doctrine* (1585). The revolt of the 'Godly' against 'approved' and established authorities in philosophy as well as in theology is well illustrated in R.B.'s book. For a later example, see Hezekiah Woodward's pedagogical assumptions: C. Webster, *The Great Instauration: Science, Medicine and Reform, 1626–1660* (1975), pp. 106–7.
47 *A dialogue, concerning the strife of our churche* (1584), p. 50.
48 *An Admonition* (1589), ed. W. Goode (1850), p. 91. There are considerable parallels between Cooper's *Admonition* and Bacon's 'Advertisement Touching the Controversies' of a few months later.
49 E.g., R. Bancroft, *Daungerous positions and proceedings, published and practised within this iland of Brytaine* (1593).
50 *Letters*, I, p. 83; Bacon is quoting James 3:15. Richard Hooker made a similar argument in *Of the lawes of ecclesiasticall politie* III (1593), 8:4,

a number there are who think they cannot admire as they ought the power and authority of the Word of God if in things divine they should attribute any force to man's Reason: for which cause they never use Reason so willingly as to disgrace Reason.

51 24 Henry VIII, c. 12: 'An Act in restraint of Appeals', in G.R. Elton, *Tudor Constitution*, p. 353; my emphasis. Shakespeare's Ulysses spoke for the Tudor ruling elite: 'O, when degree is shak'd / Which is the ladder of all high designs,/ The enterprise is sick./ How could communities ... But by degree stand in authentic peace?' *Troilus and Cressida*, I, iii, lines 101–8.

52 P. Collinson, *Puritan Movement*, pp. 385–431.

53 See L. Stone, *An Elizabethan: Sir Horatio Palavicino* (1956), pp. 248–55.

54 *Letters*, I, pp. 108–9; my interpolations.

55 Bacon kept his word in this. The notorious episode of two years later, when Essex expended much political capital attempting to persuade the queen to appoint Bacon as the Attorney-General, provides no counter-evidence: neither Burghley nor Cecil had advanced a candidate of their own.

56 Bacon used a complex allusion here. The 'contemplative planet' is not the moon, but Saturn, associated with the 'melancholic temperament': 'The humour of great men, great thinkers, prophets and religious seers. To be melancholy was a sign of genius' (F.A. Yates, *The Occult Philosophy in the Elizabethan Age* (1979), p. 51); and see R. Klibansky, E. Panofsky and F. Saxl, *Saturn and Melancholy* (1964). One is tempted to understand 'Sol' and 'Jupiter' as referring to Essex and Cecil respectively.

57 See the *Oxford English Dictionary* and the *Oxford Latin Dictionary*; the first instance of 'province' in the figurative use common today was in 1709. Thomas Elyot's *Dictionary* (1548 edition) defined 'province' as follows:

> 'Provincie' were countrayes, which the Romans gate farre from Rome, where only their officers did rule. Provincia is sometyme taken for the rule or autoritye of an officer: also an office, alsoo for a countraye or royaulme.

58 'Advertisement Touching the Controversies', *Letters*, I, pp. 93, 90.

59 B.L. Harley MS 6797 fo. 47–47v, which is the basis of the text in *Letters*, I, pp. 119–43, is corrupt. I follow Spedding's separate and later edition (1870).

60 Famous 'Mirrors' include: Erasmus' *Institutio principis Christiani* (1514); De Guevara's *The Diall of Princes* (1529: trans. 1557); Budé's *De l'institution du prince* (1547). See Q. Skinner, *Foundations*, I, pp. 213–19.

61 Elyot's *Governour* was reprinted in 1537, 1544, 1546, 1553, 1557, 1565 and 1580; Castiglione's *Il cortegiano* (*The Courtyer*) was published (in Latin) in England in 1571, 1577, 1585, 1593 and 1603, and (in Hoby's English translation) in 1561, 1577, 1588 and 1603. On 'fortitude' (often termed 'Magnificence'), see *Governour*, III, Chapters 8–10 and 14. On 'love' (as it is used by Bacon here), see *Courtyer*, IV (Peter Bembo's speeches).

62 G.R. Elton, *Reform and Renewal*, p. 14 and 14n.

63 'Fortitude' traditionally was praised as the only 'active' virtue (see *Governour*, III, 8) and 'love' as the most 'powerful' affection (see *Courtyer*, IV). Bacon was advocating a natural philosophy which was both 'active' and 'powerful'.

64 And see 'Advertisement Touching the Controversies': 'few follow the things themselves, more the names of things, and most the names of their masters' (*Letters*, I, p. 82).

65 'The contemplative planet carrieth me away wholly' (letter to Burghley, 1592).

66 P. Rossi, *Philosophy, Technology and the Arts in the Early Modern Era* (1970), provides an excellent survey of such claims; Bacon repeatedly denied them: see *Works*, III, pp. 496, 538, 611.

67 In attendance on that night were seven Privy Councillors: Sir Francis Puckering (the Lord Keeper), Burghley, Howard of Effingham, Buckhurst, Essex, Sir Thomas Heneage and Sir Robert Cecil. Also the earls of Shrewsbury, Cumberland, Northumberland, and Southampton, and the lords Windsor, Mountjoy, Sheffield, Compton, Rich, and Monteagle. For the details of this evening's complete entertainment, and the entire revels, see D. Bland (ed.), *Gesta Grayorum* (1968).

68 *Letters*, I, pp. 332–42; see also D. Bland, *Gesta Grayorum*, pp. 44–56.

4. Law

1 See S.F.C. Milsom, *Historical Foundations of the Common Law* (1981), pp. 1–59; and *The Legal Framework of English Feudalism* (1976), esp. Chapters 2 and 3; H.G. Richardson and G.O. Sayles, *Law and Legislation from Aethelbert to Magna Carta* (1966); R.V. Turner, *The English Judiciary in the Age of Glanvill and Bracton, c. 1176–1239* (1985); and P. Stein, *Legal Institutions: The Development of Dispute Settlement* (1984), esp. Part I. 'Proprietary justice and feudal government were in general harnessed by the royal power rather than opposed to it . . . and the making of the common law was largely a process of transfer to new central courts' (S.F.C. Milsom, *Foundations*, p. 16).

2 P.R. Hyams, 'Trial by Ordeal: The Key to Proof in the Early Common Law', in M.S. Arnold et al. (eds.), *On the Laws and Customs of England: Essays in Honor of Samuel E. Thorne* (1981), pp. 90–126; and R.C. Van Caenegem, *The Birth of the Common Law* (1973), Chapter 3.

3 P.E. Schramm, *A History of the English Coronation* (1937); E.H. Kantorowicz, *The King's Two Bodies: A Study in Mediaeval Political Theology* (1957), esp. Chapter 4; and F. Pollock and F.W. Maitland, *The History of English Law Before the Time of Edward I* [1898], ed. S.F.C. Milsom (1968), I, pp. 511–26.

4 S.F.C. Milsom, *Foundations*, p. 5.

5 C. Rawcliffe, 'The Great Lord as Peacekeeper: Arbitration by English Noblemen and their Councils in the Later Middle Ages', in J.A. Guy and H.G. Beale (eds.), *Law and Social Change in British History* (1984),

pp. 34–54; and C.B. Herrup, *The Common Peace: Participation and the Criminal Law in Seventeenth-Century England* (1987).

6 S.F.C. Milsom, *Foundations*, p. 11.

7 E.W. Ives, *The Common Lawyers of Pre-Reformation England: Thomas Kebell: A Case Study* (1983), p. 194.

8 William Fulbeck, *Direction, or Preparative to the study of the lawe* (1600), p. 94.

9 F. Pollock and F.W. Maitland, *History*, II, pp. 616–50; T.A. Green, *Verdict according to Conscience: Perspectives on the English Criminal Trial Jury, 1200–1800* (1986).

10 The first printed use of the word was in 1539, in the *Act of Proclamations* (31 Henry VIII, c. 8), which was written by Thomas Cromwell: '[he] shall be adjudged a traytour, and his facte high treason' (*Statutes of the Realm*, III, p. 727).

11 'Legal development consists in the increasingly detailed consideration of facts', S.F.C. Milsom, 'Law and Fact in Legal Development', *University of Toronto Law Journal*, 17 (1967), p. 1. 'What takes a legal system beyond the mere classification of claims is the adoption of a mode of trial which allows the facts to come out', Milsom (ed.), F. Pollock and F.W. Maitland, *History*, p. lxvii.

12 During the pleading of a case, some allegations might be accepted by both parties as being true; for the purposes of the suit, the court now held them to be facts.

13 Documents had the same logical status as witnesses. A document functioned as a surrogate for testimony; it preserved knowledge of an action across time and place, and just like the statements of a witness currently alive, a document's integrity was open to scrutiny.

14 W.J. Jones, *The Elizabethan Court of Chancery* (1967), pp. 249n., 252, 254n., 259.

15 On Chancery examination procedures, see W.J. Jones, *Chancery*, pp. 236–63. Examinations 'in the country' would await the conclusion of a second phase of pleading (i.e., the plaintiff's 'replication' and the defendant's 'rejoinder').

16 'Sufficiency' meant that interrogatories must be devoted to the statements in the pleadings, and to nothing else: W.J. Jones, *Chancery*, pp. 243–4.

17 P.R.O., Chancery Town Depositions: C 24/293/64 ('Pulter v. Bristowe'): quoted by W.J. Jones, *Chancery*, p. 238. The status of legal evidence was much discussed by classical and medieval logicians well known in the sixteenth century; see A. Giuliani, 'The Influence of Rhetoric on the Law of Evidence and Pleading', *Juridical Review*, 7 (1962), pp. 216–52, at pp. 230–7.

18 W.J. Jones, *Chancery*, pp. 244, 252–4.

19 There were exceptions and additions to the structure of examinations outlined here, which befitted the Chancellor's discretionary powers and equitable jurisdiction; principal among these were examinations *ad*

informandum conscientiam iudicis and *in perpetuam rei memoriam*; W.J. Jones, *Chancery*, pp. 250–2, 254–62.

20 'This Court ... doth keep all England in quiet', Coke, 4 *Inst.*, p. 65. On Star Chamber, see G.R. Elton, *Tudor Constitution*, pp. 163–87, and T.G. Barnes, 'Due Process and Slow Process in the Late Elizabethan and Early Stuart Star Chamber', *American Journal of Legal History*, 6 (1962), pp. 221–49, 315–46.

21 J. Hawarde, *Les Reportes del Cases in Camera Stellata*, ed. W.P. Baildon (1894), p. 78. The Attorney-General's *ore tenus* procedure illustrates the desire of the court to proceed quickly and firmly. If during an examination on behalf of the law officers, a defendant admitted to any matter which was a crime punishable by Star Chamber, the Attorney-General could proceed against him *ore tenus* (orally to the Court, without initial bill and so on), bypassing the time-consuming usual procedures.

22 See J.G. Bellamy, *The Tudor Law of Treason: An Introduction* (1979); G.R. Elton, *Policy and Police: The Enforcement of the Reformation in the Age of Thomas Cromwell* (1972); and J.G. Bellamy, *Criminal Law and Society in Late Medieval and Tudor England* (1984).

23 For example, SP.12/239/114 (July 1591), in which an 'intelligencer' delivered to Francis Bacon and Essex the addresses of recently arrived seminary priests and Jesuits, and a variety of other information on suspicious persons.

24 The examiners, while often present, did not themselves torture. On the 'rack-masters' and the torturing, see J.G. Bellamy, *Treason*, pp. 109–21. It seems the obsessed and twisted zealot Richard Topcliffe sometimes tortured Catholic prisoners independently of their examination sessions. The only logical rationale for legal torture (if we may talk in such terms) was the breaking of a prisoner's will to resist his questioners, but it appears that Topcliffe desired as well to punish and torment Catholics; see J.G. Bellamy, *Treason*, pp. 113–16 and J. Heath, *Torture and English Law* (1982), pp. 138–9. The use of torture in these examinations cannot be taken to show a 'reception' of continental Roman-canon law inquisitorial procedures in Tudor England. The falseness of this persistent myth has been demonstrated by several historians, for example by J.H. Langbein, *Prosecuting Crime in the Renaissance: England, Germany, France* (1974).

25 *Letters*, III, p. 114. Thomas Hobbes, *Philosophicall Rudiments concerning Government and Civill Society* (1651), Chapter 2, sect. 19; this is Hobbes' translation of his *De cive* (1642).

26 This assumes the defendant made a 'plea in bar' (the usual response) rather than a 'dilatory plea'. For a discussion of tactics in pleading and of the three sorts of plea in bar (the 'general denial', 'special traverse', and 'confession and avoidance'), see J.H. Baker (ed.), *Spelman's Reports*, II, pp. 143–52; and D.W. Sutherland, 'Legal Reasoning in the Fourteenth Century: The Invention of "Color" in Pleading', in M.S. Arnold et al. (eds.), *On the Laws and Customs*, pp. 182–94.

27 Sir Thomas Smith, *De republica Anglorum* (printed posthumously in 1583), II, chapter 12, p. 94.

28 *Tenures* (*c.* 1481), section 534.

29 *The First Part of the Institutes of the Laws of England* (1628), p. 235b (known familiarly as *Coke on Littleton* and abbreviated as Co. *Litt.*). On dialectics, see A. Giuliani, 'Influence of Rhetoric', and H.J. Berman, *Law and Revolution: The Formation of the Western Legal Tradition* (1983), pp. 132–43. During the sixteenth century, while this traditional dialectic (the *ratio artificialis*) was being attacked on the Continent by men such as Peter Ramus who advocated something new (a 'natural reason'), in England 'the techniques of *ratio artificialis* continued to enjoy the confidence of the common lawyers' (A. Giuliani, 'Influence of Rhetoric', p. 237). Sir Edward Coke argued that 'causes ... are not to be decided by natural reason but by the artificial reason and judgement of law', *12* Co. *Rep.*, 63 (1607); and see *Co. Litt.*, 97b. Cf. W.R. Prest, *The Inns of Court under Elizabeth and the Early Stuarts, 1590–1640* (1972), and L.A. Knafla, 'The Influence of Continental Humanists and Jurists on English Common Law in the Renaissance', in R.J. Schoeck (ed.), *Acta Conventus Neo-Latini Bononiensis: Proceedings of the Fourth Congress of Neo-Latin Studies* (1985), pp. 61–71.

30 S.F.C. Milsom, *Studies in the History of the Common Law* (1985), p. x; 'The largest legal changes are precisely in the understanding of things, in the assumptions upon which the system rests. And since it is the point of assumptions that one does not think about them, it follows that the largest changes are never visible until they are all over' (*ibid.*, p. 209).

31 On demurrer to the evidence, and the special and general verdicts which supplanted it, see S.F.C. Milsom, *Foundations*, pp. 72–9.

32 Few moots and readings were intended for publication. For some of those manuscript copies and notes of readings which have survived, see J.H. Baker, *English Legal Manuscripts*, 2 vols. (1975, 1978), S.E. Thorne (ed.), *Readings and Moots at the Inns of Court in the Fifteenth Century*, I (1974), and S.E. Thorne and J.H. Baker (eds.), *Readings and Moots at the Inns of Court in the Fifteenth Century*, II (1990).

33 S.F.C. Milsom, *Foundations*, pp. 44–8.

34 This follows E.W. Ives, 'The Purpose and Making of the Later Year Books', *Law Quarterly Review*, 89 (1973), pp. 64–88, and J.H. Baker (ed.), *Spelman's Reports*, II, pp. 164–78. L.W. Abbott's *Law Reporting in England, 1485–1585* (1973) appeared just as Ives' article changed our understanding of the subject. Because the printed Year Books seemed so comprehensive, it had been thought an organised system for reporting cases must have existed, and because the Year Books stop in 1535, it appeared something sinister was at work. And because from mid-century onward, we have the publication of the private reports of Tudor cases by named judges and serjeants, it was thought these reports differed substantially from those which had appeared before. Yet just like their creation, the solution to these puzzles was bibliographical: 'The uniform

appearance of the year books is, indeed, very largely the creation of the printers and supremely of Richard Tottell; it was Tottell who imposed regular divisions into terms, grouped reports into reigns, made proper citation possible by standard foliation and added copious references' (E.W. Ives, 'Later Year Books', p. 76).

35 On the 'common erudition', see E.W. Ives, 'Later Year Books', pp. 67–9 and *Common Lawyers*, pp. 158–61, and J.H. Baker (ed.), *Spelman's Reports*, II, pp. 160–3.

36 J.H. Baker, *An Introduction to English Legal History* (1979), p. 172; 'authority lay in the collective mind of the profession, past and present' (E.W. Ives, *Common Lawyers*, p. 161).

37 *7 Co. Rep.*, 27a.

38 For example, in 1544 Justice Shelley remarked of a former Chief Justice that he had been not been very learned: 'jeo pose que ceux que fueront melior literate que il fuit fueront del contrarie opinion', quoted by J.H. Baker, *Spelman's Reports*, II, p. 163n. Also see Fulbeck, *Direction, or Preparative*, p. 80.

39 F.W. Maitland (ed.), *Year Books of 3 Edward II*, vol. 3 (1905), p. xiv. Several of my examples of the uses of precedents come from T. Ellis Lewis, 'The History of Precedent', *Law Quarterly Review*, 46, 47, 48 (1930–2).

40 'The work had ceased to interest a profession that had little use for thinking in terms of substantive law. And so Bracton's was the last English law-book for centuries to be written with such terms in mind', D.E.C. Yale, '"Of No Mean Authority": Some Later Uses of Bracton', in M.S. Arnold et al. (eds.), *On the Laws and Customs*, p. 383.

41 Fitzherbert, *Le Graunde Abridgement*, 'Garde', pl. 71.

42 *8 Co. Rep.*, 35a (1587); also *1 Plowd.*, 357–9 (1563).

43 J.H. Baker, *Introduction*, p. 169.

44 For example, in his *Preparative*, Chapter 9, William Fulbeck provided the student with a detailed analysis, with charts, of the first two chapters of 'Littleton'.

45 Morgan Sjt., in *1 Plowd.*, 27 (1551), 'Colthirst v. Bejushin'.

46 Littleton, *Tenures*, section 3; *ibid.*, section 648; Christopher St Germain, *A Dialogue in Englisshe, bytweene a doctour of dyvynyte, and a student in the lawes of Englande* (1532), ed. T.F.T. Plucknett and J.L. Barton (1974), 'Dialogue I', Chapter 8, maxim 4 (p. 59); Morgan Sjt., in *1 Plowd.*, 28 (1549).

47 *1 Plowd.*, 33 (1549).

48 R. Pound, 'The Maxims of Equity: I. Of Maxims Generally', *Harvard Law Review*, 34 (1921), p. 830.

49 St Germain, *Doctour and Student*, I, 8, p. 59.

50 *Co. Litt.*, 11a.

51 St Germain, *ibid.*

52 R. Pound, 'Maxims of Equity', pp. 830–1.

53 St Germain, *Doctour and Student*, I, 8, p. 57.

54 *Co. Litt.*, 67a.

55 J.H. Baker, *Introduction*, p. 158.
56 J.H. Baker, *Spelman's Reports*, II, p. 162.
57 *1 Plowd.*, 89 (1553).
58 The *Short Title Catalogue* notes nineteen editions of Littleton's *Tenures* by 1546; the Bible (both Latin and English) reached nineteen editions only by 1551.
59 *Co. Litt.*, sig. B; *10 Co. Rep.*, 'Introduction'.
60 This phrase is not Coke's alone; for instance, it is used of Littleton by W. Fulbeck, *Preparative*, p. 71.
61 'A stately volume, perhaps the best printed law book we have ever had' (T.F.T. Plucknett, *A Concise History of the Common Law* (1956), p. 263).
62 Simon Theloall, *Le digest des briefes originals* (1579); Abraham Fraunce, *The Lawiers Logike* (1588); Edward Hake, *Epieikeia, a Dialogue on Equity* (*c.* 1595); all cited by D.E.C. Yale, 'Some Later Uses of Bracton', in M.S. Arnold et al. (eds.), *On the Laws and Customs*, pp. 383–96.
63 See his tracts in L.A. Knafla, *Law and Politics in Jacobean England* (1977), pp. 217–18, 248–9, 326–7.
64 D.E.C. Yale, 'Later Uses of Bracton', p. 388.
65 J.H. Baker (ed.), *Spelman's Reports*, II, p. 155.
66 2 *Plowd.*, 470 (1575), 'Soby v. Molins'.
67 *Co. Litt.*, 293a. My emphasis.
68 *Les Commentaries* (1571), Prologue, sig. B.
69 *Co. Litt.*, 254a.
70 *7 Co. Rep.*, 4a (1608), 'Calvin's [i.e., the *Post-nati*] Case'; and see *Co. Litt.*, 293a, and *1 Co. Rep.*, 'Preface'.
71 'And Wray Chief Justice and Sir Thomas Gawdy said, that as he who is a bastard born hath no cousin, "so every case imports suspicion of its legitimation, unless it has another case which shall be as a cousin-german, to support and prove it"' (*3 Co. Rep.*, 23a (1587), 'Walker's Case').
72 4 *Co. Rep.*, 94a (1602), 'Slade v. Morley'.
73 'This was openly deliver'd by Popham Chief Justice 9 Nov. 44 Eliz. to be the resolution of all the justices of England, and this to be a precedent for all subsequent cases' (*Yelverton*, 21 (1602), 'Slade v. Morley').
74 'The highest and most binding laws are the statutes which are established by Parliament': Coke, 2 *Inst.*, Proeme; and see 4 *Inst.*, 342.
75 4 *Co. Rep.*, 93b (1602); *7 Co. Rep.*, 28a (1608).
76 *13 Co. Rep.*, 14 (1608); *8 Co. Rep.*, 97a (1609).
77 2 *Co. Rep.*, 16b (1596), 'Lane's Case'.
78 See E.H. Levi, *An Introduction to Legal Reasoning* (1949).
79 *7 Co. Rep.*, 19a (1608).
80 'Nota bene by the reporter. From this judgement and the cause of it, the reader may observe, that it is not the words of the law, but the internal sense of it that makes the law; and our law (like all others) consists of two parts, viz. of body and soul, the letter of the law is the body of the law, and the sense and reason of the law is the soul of the law' (2 *Plowd.*, 465 (1574)).

81 7 *Co. Rep.*, 28a (1608).

82 C.K. Allen, *Law in the Making* (1964), pp. 285–311, and 'Jurisprudence – What and Why?', in his *Legal Duties and Other Essays in Jurisprudence* (1931), pp. 1–27; R.W.M. Dias, 'The Empirical Approach to the Teaching of Jurisprudence', *Law Quarterly Review*, 72 (1956), pp. 557–74; 'Home Office v. Dorset Yacht Co. Ltd.', 2 *All England Reports* (1970), p. 323, *per* Lord Diplock.

83 B.E. King, 'The Concept of a Lawyer's Jurisprudence', *Cambridge Law Journal*, 11 (1952–3), pp. 229–39 and pp. 404–20.

84 R. Cross, *Precedent in English Law* (1977), pp. 180–8.

85 J. Stone, *Legal System and Lawyers' Reasonings* (1964), and his *Precedent and Law: Dynamics of Common Law Growth* (1985); D. Lloyd, 'Reason and Logic in the Common Law', *Law Quarterly Review*, 64 (1948), pp. 468–84.

86 A.G. Guest, 'Logic in the Law', in Guest (ed.), *Oxford Essays in Jurisprudence* (1961), p. 182.

87 In Easter Term 1591, Bacon argued in 'Hulme v. Jee': *Moore (K.B.)*, p. 301. In Lent 1594, he argued in 'Lord Cheyney's Case': *Moore (K.B.)*, p. 727; *5 Co. Rep.*, p. 68a. In Easter Term 1594, he argued in 'Dillon v. Frain' (i.e., 'Chudleigh's Case'): *Popham*, p. 70; *1 Anderson*, p. 309; *1 Co. Rep.*, p. 113. Bacon's argument is B.L. Lansdowne MS 1121, fos. 112–20v, translated in *Works*, VII, pp. 617–36. Spedding (*Letters*, I, p. 267) claimed Bacon had never argued a case before 1594, but this is wrong; moreover, his opponent in 'Hulme v. Jee' was none other than Edward Coke. In 1600, at Gray's Inn, Bacon read on the Statute of Uses (*Works*, VII, pp. 395–445), repeating much of his argument in 'Chudleigh's Case', emphasising the gravity of the problems unsolved (even exacerbated) by the Statute of Uses. In 1602, Bacon argued before the judges concerning the law of contracts; see J.H. Baker, 'New Light on Slade's Case', *Cambridge Law Journal*, 29 (1971), pp. 51–67, 213–36. On the importance of 'Chudleigh's Case' and 'Slade's Case' see J.H. Baker, *Introduction*, p. 240 *et passim*; S.F.C. Milsom, *Foundations*, pp. 228, 249–56; and A.W.B. Simpson, *A History of the Land Law*, pp. 218–23.

88 See *Letters*, I, pp. 232–4 *et passim*. The story does not originate with Spedding, but his presentation and explanation of this episode have been used as authoritative for more than a century.

89 *Letters*, I, pp. 267–8.

90 *A True Report of the Detestable Treason Intended by Dr Roderigo Lopez*, *Letters*, I, pp. 274–87.

91 Bacon's post, strictly speaking, was that of 'Deputy Chief Steward (South Parts)', i.e., everywhere south of the River Trent. R. Somerville blandly described the duties of the office as 'holding courts and dealing with legal technicalities', *History of the Duchy of Lancaster*, I, p. 332.

92 13 June 1594 at the Tower. SP.12/249/11 is a declaration about an interrogation conducted that day, and is signed by (among others) Bacon,

Fleming, William Lambarde, and Bacon's old acquaintances Richard Topcliffe and Richard Young.

93 See SP.12/249, item 87 (16 Aug.); items 96–8 (20 Aug.); 110 (21 Aug.); 112–14 (24 Aug.); 117–24 (27 Aug.); 125–7 (28 Aug.); 132 (31 Aug.); and SP.12/250/7 (12 Sept.). These letters may originally have been among Essex's papers: 12/249/110 implies he was the recipient of Bacon's and Waad's information, and 12/249/117 has his scrawled notes upon it. The interrogations were conducted primarily by Bacon and William Waad (Clerk of the Privy Council, a former clerk of Walsingham's, and with Bacon in Paris in the late 1570s), and they were joined occasionally by Attorney Coke, Sir Michael Blount (the Recorder of London) and Serjeant Edward Drew.

94 See B.L. MS Harley 1697, fo. 43, discussed by J.H. Baker, *The Order of Serjeants of Law* (1984), p. 112 and 112n. There was much prickly debate among the serjeants in the 1590s about their superior *status et gradus* ('estate and dignity'), but Bacon may well have achieved two valuable rights: 'preaudience', the right to have your case heard before those of ordinary barristers (see J.H. Baker, *Serjeants*, pp. 56–64) and the serjeants' exclusive right of 'addressing the court on points of law arising from cases pleaded in the court, whether before or after the trial of the issue' (J.H. Baker, *Serjeants*, p. 43). On 25 January 1595, in a letter to his brother Anthony, Bacon related a conversation with Robert Cecil in which he was told that the queen had said she 'hath pulled me over the bar' (*Letters*, I, p. 348). This phrase is commonly assumed to refer to Bacon's becoming an utter barrister in 1582; but surely that involved pulling Bacon up to the bar, not over it, and the reference is clearly and precisely about being brought 'within the bar' in 1594. For infamous conduct, solicitors were 'pushed over the bar'; J.H. Baker, 'Solicitors and the Law of Maintenance, 1590–1640' (1973), in J.H. Baker, *The Legal Profession and the Common Law: Historical Essays* (1986), p. 126, 126n.

95 At the beginning of February 1594, Bacon had appeared before the justices in King's Bench in 'Lord Cheyney's Case', which presumably had arrived there on a writ of error from the Court of Wards where it had originated in Michaelmas 1591. In Easter Term 1594, Bacon took over from Coke in 'Chudleigh's Case'. In letters of July and September 1595, Bacon mentions assisting Coke in other cases (*Letters*, I, pp. 363, 367).

96 Letters generated by these manoeuverings are printed in *Letters*, I, pp. 313–69.

97 Francis to Anthony Bacon, 25 January 1595 (*Letters*, I, p. 348). 'I have been like a piece of stuff bespoken in the shop; and if her Majesty will not take me, it may be the selling by parcels will be more gainful. For to be, as I told you, like a child following a bird, which when he is nearest flieth away and lighteth a little before, and then the child after it again, and so *in finitum*, I am weary of it' (Francis to Fulke Greville in May 1595, *Letters*, I, p. 359).

98 11 October 1595, *Letters*, I, p. 368.

99 *Sir Francis Bacon His Apologie in Certain Imputations concerning the late Earl of Essex* (1604), *Letters*, III, p. 144. And see his letter to Essex in late 1595: 'For your Lordship, I do think myself more beholding to you than to any man. And I say, I reckon myself as a *common* (not popular, but *common*); and as much as is lawful to be enclosed of a common, so much your Lordship shall be sure to have' (*Letters*, I, p. 373).

100 *Letters*, II, p. 41.

101 That Bacon had such access is clear from his correspondence printed in *Letters*, II.

102 In the Public Record Office there are ample instances of this. For 1596: see SP.12/257/66; 12/259/26, 27, 54; and 12/260/69. For 1597: 12/262/123 and 12/265/83. For 1598: 12/268/62, 83–6, 89–91, 103. For 1600: 12/275/115. For 1601: 12/279/23–4, 58; 12/281/68; 12/282/6. For 1602: 12/286/42. The *Calendar of Salisbury MSS* indicates that Bacon examined prisoners in June 1599 (vol. IX, p. 208) and July 1600 (vol. X, p. 256).

103 On these and other warrants, see J. Heath, *Torture and English Law* (1982), and J.H. Langbein, *Torture and the Law of Proof: Europe and England in the Ancien Regime* (1977). The warrants to Bacon are printed by Heath (pp. 224–7). Heath argues (p. 135) there was at least one more case in which Bacon was involved where torture was employed (at the Tower, on 19 October 1598; see SP.12/268/83); no warrant survives, but the circumstantial evidence is indeed strong (see *Cal. Salisbury MSS*, VIII, p. 382). Typically, each investigation took many months, if not years, and a torture warrant presumably would hold good against that named individual throughout the full course of the investigation. For example, a warrant to torture Valentine Thomas was addressed to Bacon et al. on 17 April 1598; they were still questioning him on 18 December (*Cal. Salisbury MSS*, VIII, p. 505).

104 These figures are Heath's, *Torture*. Langbein, *Torture*, gives figures of eighty-one in total, fifty-nine of which are Elizabethan and thirty-one from the war years. The discrepancies come from differing ways of deciding whether a warrant refers to an ongoing or a new case.

105 Heath, *Torture*, pp. 131–6, referring to contemporary Catholic chronicles, suggests there were in the 1590s several cases in which torture was used for which no record of a warrant survives. Even allowing for incomplete records, it remains the case that the greatest number of approvals of torture came during (and because of) the war.

106 My emphasis. This warrant was issued to interrogate the ringleaders of the 1596 Oxford rising. See J. Walter, '"A Rising of the People"? The Oxfordshire Rising of 1596', *Past and Present*, 107 (1985), pp. 90–143: the local magistrates and the central government feared outright rebellion.

107 D'Ewes' *Journals* tell us that in 1593 he was on committees for 'punishment of rogues' (pp. 476b–477a, 499a) and for 'reducing disloyal subjects to their true obedience' (p. 481b); in 1597, committees for the 'increase of People for the strength and service of the Realm' (pp. 565a,

570a, 578b, 581a, 583b) and for 'House of Correction and the punishment of Rogues and sturdy Beggars' (pp. 558a, 579a, 581a, 582a). Bacon had been made a JP for both Middlesex and Essex in early 1593. Rather than from any anachronistic 'constitutional' qualms, Bacon's criticism of the 1593 subsidy proposal stemmed from his security anxieties: 'We shall first breed discontentment in paying these Subsidies, and in the Cause endanger her Majesty's safety, which must consist more in the love of her people than in their wealth', (D'Ewes, p. 493). Fragments of memoranda on intelligence and the safety of the queen (written in 1594) are printed in *Letters*, I, pp. 306–8.

108 See Hawarde's *Les Reportes*, pp. 61, 78. The case in 1597 was about engrossing (hoarding grain), which in years of repeated harvest failures could easily provoke riots: an obvious case for the attention of Star Chamber and the security officers of the state.

109 These tracts are printed in *Letters*, I and II. The official character of Bacon's tract against Parsons is confirmed by the existence in the Public Record Office of a fair copy of an intended introduction to it, endorsed by Robert Cecil: SP.12/242/18. In the 1593 parliament, Bacon followed three Privy Councillors (Cecil, Fortesque, Wolley) in speaking about 'the sundry particular practices of the King of Spain against this State of this Realm': D'Ewes, *Journals*, p. 471.

110 Papers and tracts concerning Essex's trial are printed in *Letters*, II, pp. 175–365.

5. A reformed state

1 My account of Bacon's legal reform differs substantially from that of B.J. Shapiro, 'Law Reform in 17th Century England', *American Journal of Legal History*, 19 (1975), pp. 280–312; 'Sir Francis Bacon and the Mid-Seventeenth Century Movement for Law Reform', *American Journal of Legal History*, 24 (1980), pp. 331–62; and *Probability and Certainty in Seventeenth-Century England* (1983); and from that in P.G. Burgess, 'Custom, Reason and the Common Law: English Jurisprudence, 1600–1650' (unpublished Ph.D. thesis, Cambridge University, 1988), pp. 136–41.

2 *Letters*, I, pp. 339–40. This was a theme dear to his father (see Chapter 1, nn. 52, 57).

3 *Letters*, I, p. 214; Simonds D'Ewes, *Journals* (1682), p. 473.

4 *Letters*, II, p. 79; Hayward Townshend, *Historical Collections* (1680), p. 103. In 1588, Bacon had been involved in such a review (see Chapter 2, n. 47).

5 *Letters*, III, p. 19; Townshend, *Collections*, p. 184. Bacon continued to repeat this opinion; see *Works*, III, p. 475 (1605); *Letters*, III, p. 336 (1607); *Letters*, V, p. 41 (1614); *Letters*, VII, pp. 177–8 (1620); *De augmentis scientiarum* (1623), VIII, *Aphorismus* 8, *Works*, I, p. 805.

6 *Letters*, I, p. 214; D'Ewes, *Journals*, p. 473. He repeated this in the 1606–7 parliament; see D.H. Willson (ed.), *Parliamentary Diary of Robert Bowyer* (1931), p. 268, and *Commons Journals*, I (1800), p. 1037. See also *Letters*, VI, p. 66 (1616); and *Letters*, VII, pp. 360–2 (1622).

7 *Letters*, III, p. 19; and see *Letters*, V, pp. 15, 41 (1614).

8 *Letters*, III, p. 230; and see *ibid.*, p. 336 (1607).

9 *Letters*, V, pp. 84–6; and see *Letters*, VI, pp. 57–71 (1616); *Letters*, VII, pp. 358–64 (1622); *Works*, I, pp. 817–18: Aphorismi 59–64 (1623).

10 And see *Letters*, VI, p. 69; *Letters*, VI, pp. 265–6; *Works*, I, p. 821: Aphorismi 73–6. When he became Lord Keeper, Bacon instituted official reporters in his Court of Chancery.

11 *Letters*, VI, p. 68.

12 SP.14/28/67 (printed in *Letters*, III, pp. 389–92). The proclamation as published closely followed this wording; see Larkin and Hughes (eds.), *Stuart Royal Proclamations* (1973), I, pp. 167–71 (proclamation no. 77).

13 Letter to George Villiers; two drafts are printed in *Letters*, VI, pp. 13–26, 27–56.

14 He further remarked: 'The House of Commons hath only power to censure the members of their own house, in point of election and misdemeanors in or towards that house; and have not, nor ever had the power so much as to administer an oath in any case whatsoever to prepare a judgement' (*Letters*, VI, p. 38).

15 *Letters*, IV, p. 177; and see *Works*, VII, p. 370 (*Maxims of the Law*, *regula* 19). According to Ellesmere, 'the Popular state [the Commons], ever since the begynnynge of his majesties gracious and sweete government, hath growne bygge and audacious. And in every session of Parlement swelled more and more. And yf waye be stylle gyven unto yt (as of late hath bene), yt is to be Doubted what the ende wyll be': 'Speciall Observations Touching the ... Last Parliament' (1611), in L.A. Knafla, *Law and Politics*, pp. 254–62 at p. 254.

16 *Letters*, VI, p. 34; and see Bacon's speech to the judges in 1617: 'You must remember, that beside your ordinary administration of justice, you do carry the two glasses or mirrors of the state; for it is your duty in these your visitations to represent to the people the graces and care of the King: and again, upon your return, to present to the King the distastes and griefs of the people', *Letters*, VI, p. 211; see also *Letters*, V, p. 143 (1615).

17 *Letters*, VI, p. 192.

18 *Ibid.*, p. 211.

19 *Ibid.*, p. 303.

20 *Ibid.*, p. 68.

21 Bacon continued: 'Towards this there hath been already, upon my motion and your Majesty's direction, a great deal of good pains taken; by my Lord Hobart, myself, Serjeant Finch, Mr. Heneage Finch, Mr. Noye, Mr. Hackwell, and others' (*ibid.*, p. 71).

22 That is, 'a preparation to the reading of the corps of the common law'; Bacon was not speaking of a 'course of study' for pupils.

23 A version of the manuscript presented to the queen is printed in *Works*, VII, pp. 301–87, but this is D.D. Heath's polyglot copy-text; it has no authority. Yet Heath's suggestions about the composition and dating seem probable: the valuable point being that Bacon prepared drafts between *c.* 1593 and the Christmas season 1596–7 (drafts which are best represented among the eleven extant manuscripts by C.U.L. MS Ff.4.16). I am presently preparing this and other Elizabethan texts for a new critical edition of Bacon's works.

24 *Works*, VII, p. 320; and see *ibid.*, p. 333 (*regula* 3).

25 Bacon's proposals depended upon a strict following of precedents. He argued for this in a Star Chamber case in June 1614: 'All the law of England in effect is precedent law, that is to say, when the judges upon deliberation have resolved and adjudged the controversy, this judgement and resolution is precedent to all posterity afterwards', Folger MS V.a.133, fo. 35; quoted by T.G. Barnes, 'A Cheshire Seductress, Precedent, and a "Sore Blow" to Star Chamber', in M.S. Arnold et al. (eds.), *On the Laws and Customs*, pp. 359–82 at p. 370.

26 SP.14/28/67 (*Letters*, III, pp. 391–2).

27 This concern over jurors was not new; in 1590, an office had been established to collect fines from freemen defaulting on jury duty; W. Notestein et al. (eds.), *Commons Debates, 1621*, VII, p. 387.

28 *Letters*, VI, p. 37. And see his father's remarks to the same effect, quoted in Chapter 1, p. 20.

29 *Ibid.*, p. 34. King James made similar pronouncements; see W.S. Holdsworth, *History of English Law*, VI, pp. 57–8.

30 'Of Judicature', in *The Essaies* (1612 edition), *Works*, VI, p. 585.

31 'Of Judicature', in *The Essayes* (1625 edition), ed. M. Kiernan (1985), p. 169.

32 *Letters*, VI, p. 211 (1617); *ibid.*, p. 303 (1618).

33 *The Essaies* (1612); see also *Letters*, VI, p. 202 (1617). Ellesmere remarked in *c.* 1604 of some 'subordinate Judges' that 'it may seeme they rather desire to be Kings, than to rule the people under him, whoe will not administrate Justice by lawe but by therie owne willes': 'The Royall Prerogative', in L.A. Knafla, *Law and Politics*, pp. 197–201 at p. 199. And see King James' speech to the judges in Star Chamber in June 1616; *The Political Works of James I* (1616), ed. C.H. McIlwain (1918), p. 326.

34 *Letters*, V, pp. 84–6; elsewhere Bacon spoke of 'seminaries of doubts and incertainties': *Letters*, VI, p. 69. His alarm about 'voluntary communities' and religious sects carried directly into alarm about 'voluntaries' and 'seminaries of doubts' among the lawyers. Furthermore, in 1617, Sir Julius Caesar took notes on a speech by Bacon to the judges, in which Bacon emphasised his concerns about the dangers of unmediated knowledge: 'New opinions spread very dangerous, the late Traske a dangerous person. Prentices learn the Hebrew tongue' (*Letters*, VI, p. 315).

35 See *Letters*, III, pp. 1–5.

36 The story of Coke, King James and the royal prerogatives has been told

repeatedly, most often by historians whose central concerns are the Civil War and Revolution. For example, see S.R. Gardiner, *The History of England* (1884); C.H. McIlwain, *The High Court of Parliament and Its Supremacy* (1910); W. Notestein, *The Winning of the Initiative by the House of Commons* (1924); M. Judson, *The Crisis of the Constitution* (1949); G. Mosse, *The Struggle for Sovereignty in England* (1950); S.D. White, *Sir Edward Coke and the 'Grievances of the Commonwealth'* (1979). Cf. G.R. Elton, 'A High Road to Civil War?' (1965), in *Studies*, II, pp. 164–82; C. Russell, 'Parliamentary History in Perspective, 1604–1629', *History*, 61 (1976), pp. 1–27; P. Christianson, 'The Causes of the English Revolution: A Reappraisal', *Journal of British Studies*, 15 (1976), pp. 40–75; and M. Kishlansky, 'The Emergence of Adversary Politics in the Long Parliament', *Journal of Modern History*, 49 (1977), pp. 617–40.

37 *Letters*, IV, pp. 379, 382.

38 Consider Bacon's memorandum of 1613 about how to manage the next parliament; for example: 'What is fit to be done for the winning or bridling of the Lawyers (which are the *literae vocales* of the house) that they may furder the King's causes, or at least fear to oppose them' (*Letters*, IV, pp. 366–7).

39 See, for example, S.R. Gardiner, *History of England*, III, *passim*, and C.D. Bowen, *The Lion and the Throne: The Life and Times of Sir Edward Coke, 1552–1634* (1957), pp. 378–95.

40 *Letters*, VI, pp. 80, 86–96. And see Ellesmere's 'Observacions upon Cooke's Reportes' (1615), in L.A. Knafla, *Law and Politics*, pp. 297–318.

41 *Letters*, VI, p. 95. Ellesmere concurred: 'In all his Cases sett forth by him since he was a Judge every Pattent is made good whereby the King parteth with his Inheritance, and every Pattent is made void by which his Majestie would express his Power' (L.A. Knafla, *Law and Politics*, p. 302).

42 'Of Judicature' (1612), *Works*, VI, p. 582; and see *Letters*, V, pp. 252, 253. Ellesmere argued: 'Wee have in this age so many Questionists; and *Quo modo*, and *Quare*, are so common in most mens mouthes, that they leave neither Religion, nor Lawe, nor King nor Counsell, nor Policie, nor Government out of question' ('Touching the Post-nati', in L.A. Knafla, *Law and Politics*, pp. 202–53 at p. 216).

43 *Works*, VII, p. 320.

44 *Letters*, VII, pp. 351–2.

45 *Letters*, I, p. 214; quoted above, p. 107.

46 C.U.L. MS Ff.4.16, fos. 5–5v; and see *Works*, VII, p. 314. Lorenzo Valla and others made a similar criticism of Justinian: see P. Stein, *Regulae Iuris: From Juristic Rules to Legal Maxims* (1966), p. 163.

47 For example, in his letter to King James in 1616 about his plans to reform the laws, Bacon provided a most telling series of potential (and accurate) criticisms of his own proposals with careful responses to them; *Letters*, VI, pp. 63–7.

48 See B.P. Levack, *The Civil Lawyers in England, 1603–1642: A Political Study* (1973), *passim*.

49 See J. Langbein, *Prosecuting Crime* (1974). Certain English courts (that of the Admiralty and all the Church's courts) used civil law process at this time, and consider the huge rows over the High Commission's use of the *ex officio* oath – a civilian method; see M.H. Maguire, 'Attack of the Common Lawyers on the Oath *Ex Officio* as Administered in the Ecclesiastical Courts of England', in *Essays in History and Political Theory in Honor of Charles Howard McIlwain* (1936), pp. 199–229.

50 Dr John Cowell's *The Interpreter*; see Levack, *Civil Lawyers*, *passim*. In his speech during the 'Post-nati' case (1608), Ellesmere made a point of praising the civil law; L.A. Knafla, *Law and Politics*, p. 221.

51 *Corpus iuris civilis: digesta*, Book L, titles xvi, xvii; see P. Stein, *Regulae Iuris* on the fertility of the title *De regulis* for subsequent elaboration and growth in the science of the civil law.

52 The civil law was not monolithic; it made ample allowance for local custom and variations in process, trial procedure and laws. See D.R. Kelley, 'History, English Law and the Renaissance', *Past and Present*, 65 (1974), pp. 24–51; and P. Stein, *Legal Institutions* (1984).

53 See R.C. Van Caenegem, *Judges, Legislators and Professors: Chapters in European Legal History* (1987), p. 152.

54 *Letters*, III, p. 26 (1601). In this, and in many sentiments about monarchy, Bacon's convictions closely match those of King James, in his *The Trew Law of Free Monarchies* (1598), and *A Speech to the Lords and Commons of the Parliament* (1610); see *The Political Works*, ed. C.H. McIlwain (1918).

55 These three passages are *Letters*, III, pp. 371, 371–2 (see SP.14/10/86 and 87), and p. 373; and see *Works*, VII, pp. 587–611.

56 *Works*, VII, p. 647 (1608).

57 *Letters*, IV, p. 350; see also Bacon's charge in Star Chamber against Oliver St John: *Letters*, V, pp. 136–46 (1615).

58 'Aphorismus 12', in 'Aphorismi de jure gentium maiore sive de fontibus justiciae et juris', in Hardwick MS 51.11. This Latin text has been printed by M. Neustadt, 'The Making of the Instauration' (unpublished Ph.D. thesis, The Johns Hopkins University, 1987), pp. 239–71. I have made some alterations in his translation. Neustadt argues for the close relationship of this tract to a section of *De augmentis*, VIII (1623), but dates 'Aphorismi' to early in James' reign; I see no evidence for this dating, and place it nearer the time of the writing of *De augmentis*.

59 *Letters*, III, p. 63; see also *ibid.*, pp. 68, 235 (1604).

60 *Letters*, III, p. 69. In 1602 (*ibid.*, pp. 45–51), Bacon had laid out a plan for 'reducing that kingdom to civility'. Among his prescriptions were arrrangements for governance ('in every principal town or place of habitation ... a captain or a governor, and a judge ... who may have a prerogative commission to hear and determine *secundum sanam discretionem*') and the 'undertakers' were to proceed 'according to a prescript or a formulary' – no 'voluntaries' were to be allowed. See also *Letters*, III, p. 314 (1603); *Letters*, IV, p. 74 (1608); and *Letters*, VI, p. 206 (1617).

61 *Letters*, III, pp. 94–5. See 'Of the True Greatness of the Kingdom of Britain' (*c.* 1608), *Works*, VII, pp. 47–64; and 'Of the Greatness of Kingdomes', *The Essayes* (1625): 'All States, that are liberall of Naturalization towards Strangers, are fit for Empire', M. Kiernan (ed.), p. 94.

62 *Letters*, III, p. 323; and see pp. 324–5. See also *Letters*, VI, pp. 49–52 (1616). Bacon's downplaying of arguments from present material utility distinguishes his perspective on colonisation from that of other men once patronised by Walsingham, such as the Hakluyts, whose bellicose plans found no royal favour; see K.R. Andrews, *Trade, Plunder and Settlement*, pp. 10–14.

63 *Letters*, VII, pp. 25, 28.

64 *The Essaies* (1612), *Works*, VI, p. 588.

65 Pindar, *Olympic Ode*, III.

66 This is not to suggest Bacon wrote *New Atlantis* after the publication of *Instauratio magna*; we do not know when he wrote the surviving text, let alone when he first conceived of it. For example, his diaries of *c.* 1608 contain notes about exactly such a natural philosophical community: *Letters*, IV, pp. 66–7.

67 See, for example, F.E and F.P. Manuel, *Utopian Thought in the Western World* (1979), pp. 243–60; and P.J. Zetterburg, 'Echoes of Nature in Solomon's House', *Journal of the History of Ideas*, 43 (1982), pp. 179–93. My text is *Works*, III, pp. 129–66.

68 See his chaplain's prefatory note to the posthumously printed text.

69 A highly bureaucratised elaboration upon the well-known technological 'intelligence' practices of Lord Burghley; see especially vol. 101 of the B.L. Lansdowne MSS (and Chapter 1, n. 48).

70 *Timaeus*, 25B–C.

6. A reformed natural philosophy

1 See 'In Praise of Knowledge' (1592), *Letters*, I, p. 123; 'Filum labyrinthi, sive formula inquisitionis' (*c.* 1606), *Works*, III, p. 499; 'Praefatio', *Instauratio magna* (1620), *Works*, I, p. 132 (IV, p. 21); *Novum Organum* (1620), I, Aph. 81, *Works*, I, p. 188 (IV, p. 79).

2 'Valerius Terminus: Of the Interpretation of Nature' (*c.* 1602), *Works*, III, p. 222. Also see 'Filum labyrinthi', *Works*, III, p. 501.

3 *Meditationes sacrae* (1597), *Works*, VII, p. 252. See also 'Valerius Terminus', *Works*, III, p. 221; *Advancement of Learning* (1605), *Works*, III, pp. 267–8; and *The Essaies* (1612), *Works*, VI, p. 559.

4 *Essaies* (1612), *Works*, VI, p. 559. On sixteenth- and seventeenth-century atomists, see: R. Kargon, *Atomism in England: Hariot to Newton* (1966); L.S. Joy, *Gassendi the Atomist: Advocate of History in an Age of Science* (1988).

5 *Essaies* (1612), *Works*, VI, pp. 559–60.

6 *Letters*, III, pp. 90–1. Bacon is *not* drawing an analogy between political

laws and 'natural law' (a much-championed ethical argument, particularly in its formulations by Aquinas and the civil jurists); he is alluding to 'laws', hidden structures, in nature.

7 *Letters*, I, p. 334.

8 *Works*, III, p. 269. 'And as for Virgil's verses, though it pleased him to brave the world in taking to the Romans the art of empire, and leaving to others the arts of subjects; yet so much is manifest, that the Romans never ascended to that height of empire till the time they had ascended to the height of other arts', *Advancement of Learning*, *Works*, III, pp. 273–4. Bacon refers here to Virgil, *Aeneid*, VI, 851–3.

9 *Works*, III, pp. 348–9. During the parliamentary debates about the *post-nati* in 1607, Bacon argued for a general naturalisation, and a union of the laws of the two kingdoms, resting his case upon his belief in the 'great affinity' between 'the administration of the world under the great monarch, God himself' and that of civil states; *Letters*, III, p. 314.

10 *Advancement of Learning*, *Works*, III, p. 299, and also pp. 332, 337. Also see 'Valerius Terminus', *Works*, III, p. 220; 'Filum labyrinthi', *Works*, III, p. 500; 'Cogitata et visa' (1607), *Works*, III, p. 610; 'Praefatio', *Instauratio magna* (1620), *Works*, I, p. 132 (IV, p. 20); *Novum Organum* (1620), I, Aph. 129, *Works*, I, p. 221 (IV, p. 114). The royal 'commandment of wits and means' was, of course, Bacon's repeatedly stated desire from 1592 onward.

11 'Praefatio', *Instauratio magna*, *Works*, I, p. 132 (IV, p. 20).

12 'Prooemium', *ibid.*, p. 121 (IV, p. 7).

13 *Novum Organum*, I, Aph. 124, *Works*, I, p. 218 (IV, p. 110).

14 [*Ipsa scientia potestas est*]: *Meditationes sacrae* (1597), *Works*, p. 241. See also 'Cogitata et visa' (1607), *Works*, III, p. 611; 'Distributio operis', *Instauratio magna*, *Works*, I, p. 144 (IV, p. 32); and *Novum Organum*, I, Aph. 3: 'Human knowledge and human power meet in one; for where the cause is not known the effect cannot be produced. Nature to be commanded must be obeyed; and that which in contemplation is as the cause is in operation as the *rule*' [*instar regulae est*], *Works*, I, p. 157 (IV, p. 47); my emphasis.

15 *Letters*, III, p. 87.

16 These include: 'Cogitationes de natura rerum' (*c.* 1603), *Works*, III, pp. 15–35 (V, pp. 419–39); 'De interpretatione naturae sententiae xii' (*c.* 1606), *Works*, III, pp. 785–8; 'Partis instaurationis secundae delineatio et argumentum' (*c.* 1606), *Works*, III, pp. 547–85; 'Filum labyrinthi' (*c.* 1606), *Works*, III, pp. 496–504; 'Cogitata et visa' (*c.* 1607), *Works*, III, pp. 591–620; 'Prodromi, sive anticipationes philosophiae secundae' (*c.* 1607), *Works*, II, pp. 690–2; 'Scala intellectus, sive filum labyrinthi' (*c.* 1607), *Works*, II, pp. 687–9; 'Inquisitio du motu', 'De calore et frigore', and 'De sono' (*c.* 1607), *Works*, III, pp. 625–80; 'Thema coeli' (1612), *Works*, III, pp. 769–80; 'De fluxu et refluxu maris' (*c.* 1616), *Works*, III, pp. 47–61; 'De principiis atque originibus' (*c.* 1618), *Works*, III, pp. 79–118.

17 'Distributio operis', *Works*, I, p. 134 (IV, p. 22):
Partitiones Scientiarum.
Novum Organum, sive Indicia de Interpretatione Naturae.
Phenomena Universi, sive Historia Naturalis et Experimentalis ad condendam Philosophiam.
Scala Intellectus.
Prodromi, sive Anticipationes Philosophiae Secundae.
Philosophia Secunda, sive Scientia Activa.

18 *Ibid.*, p. 135 (IV, pp. 22–3). And consider: 'I have taken all knowledge to be my province' (*Letters*, I, p. 109); and Bacon to Robert Cecil in 1605, *Letters*, III, p. 253.

19 'Distributio operis', *Instauratio magna*, *Works*, I, p. 137 (IV, p. 25); and see *Novum Organum*, I, Aph. 105, *Works*, I, p. 205 (IV, p. 97); and *Advancement of Learning*, *Works*, III, p. 387.

20 *Letters*, I, p. 82.

21 *Works*, III, pp. 241–2, and on pages 242–52 Bacon elaborates this declaration. Idols of the 'Palace' were subsequently called Idols of the 'Marketplace'. For further discussion of the four Idols, see *Novum Organum*, I, Aphorismi 39–44, *Works*, I, pp. 163–5 (IV, pp. 53–5).

22 *Works*, III, p. 244.

23 *Advancement of Learning*, *Works*, III, p. 265. My emphasis.

24 'Distributio operis', *Instauratio magna*, *Works*, I, p. 139 (IV, p. 27). Book I of *Novum Organum*, for example, is devoted to these three 'refutations' (esp. see Aph. 115). The Idols were very deeply imbedded indeed: 'It has been well observed that the follies and superstitions which nurses instil into children do serious injury to their minds' ('Distributio operis', *ibid.*, I, p. 142 (IV, p. 30).

25 'Praefatio', *Instauratio magna*, *Works*, I, p. 130 (IV, p. 19).

26 'Praefatio', *Novum Organum*, *Works*, I, p. 154 (IV, p. 43).

27 'Distributio operis', *Instauratio magna*, *Works*, I, p. 134 (IV, p. 22).

28 'Praefatio', *Instauratio magna*, *Works*, I, p. 129 (IV, p. 18). 'For my way of discovering the sciences goes far to level wits, and leaves little to individual excellence: because it settles everything by the surest rules and demonstrations', *Novum Organum*, I, Aph. 122, *Works*, I, p. 217 (IV, p. 109).

29 'Praefatio', *Novum Organum*, *Works*, I, pp. 151–2 (IV, p. 40).

30 *Novum Organum*, II, Aph. 10, *Works*, I, pp. 235–6 (IV, p. 127); see *Works*, III, p. 187, and *Works*, V, p. 131.

31 *Novum Organum*, I, Aph. 95, *Works*, I, p. 201 (IV, p. 93); my emphasis. And see 'Cogitata et visa', *Works*, III, p. 616; and 'Descriptio globis intellectualis', *Works*, III, p. 731 (V, pp. 507–8).

32 'Distributio operis', *Instauratio magna*, *Works*, I, pp. 136–7 (IV, p. 25); 'Partis instaurationis secundae delineatio et argumentum', *Works*, III, pp. 547–8; and see *Novum Organum*, I, Aphorismi 69, 82, *Works*, I, pp. 179, 190 (IV, pp. 70, 81).

33 *Letters*, III, p. 86 (letter of 1603); *Advancement of Learning*, *Works*, III,

pp. 351–2; and *Novum Organum*, I, Aph. 70, *Works*, I, pp. 179–80 (IV, p. 70).

34 *Advancement of Learning*, *Works*, III, p. 330.

35 'Descriptio globis intellectualis', *Works*, III, p. 731 (V, p. 505).

36 *Ibid.*, III, p. 732 (V, p. 506); and see *Advancement of Learning*, *Works*, III, p. 333; and 'Phaenomena universi', *Works*, III, p. 688.

37 'Distributio operis', *Instauratio magna*, *Works*, I, p. 141 (IV, p. 29); *Novum Organum*, I, Aph. 98, *Works*, I, p. 203 (IV, p. 95).

38 *Parasceve ad historiam naturalem et experimentalem*, *Works*, I, pp. 398–9 (IV, p. 257); *Advancement of Learning*, *Works*, III, pp. 332–3.

39 *Parasceve*, *Works*, I, pp. 396–7 (IV, pp. 254–5).

40 *Ibid.*, *Works*, I, p. 401 (IV, pp. 259–60).

41 *Ibid.*, *Works*, I, p. 402 (IV, p. 261); see 'Distributio operis', *Instauratio magna*, *Works*, I, pp. 142–3 (IV, p. 30); 'Phaenomena universi', *Works*, III, p. 690: 'Cum vero ... exquisitas'.

42 *Novum Organum*, I, Aph. 100, *Works*, I, p. 203 (IV, p. 95). 'Just as if some kingdom or state were to guide its counsels and affairs, not by letters and reports from ambassadors and trusted messengers, but by town gossip and the gutters; such exactly is the administration established in philosophy with respect to Experience' (*Novum Organum*, I, Aph. 98, *Works*, I, p. 202 (IV, p. 94)).

43 'Distributio operis', *Instauratio magna*, *Works*, I, pp. 138–9 (IV, p. 26); and see *Novum Organum*, I, Aph. 70, *Works*, I, pp. 179–80 (IV, p. 70): 'Sed demonstratio ... fallax est'; and 'Phaenomena universi', *Works*, III, p. 688.

44 *Advancement of Learning*, *Works*, III, p. 363; my emphasis.

45 *Ibid.*, p. 354.

46 *Ibid.*, pp. 356–7. The quotation is Ecclesiastes 3:11: 'The works which God worketh from beginning to end'.

47 'Cogitata et visa', *Works*, III, pp. 617–18; I have incorporated some of the phrases in the so-called Gruter text of 'Cogitata et visa', as recorded in the footnotes of *Works*, I; these are not significant alterations but additional phrases, which help the sense of the passage. See also *Novum Organum*, I, Aph. 102, *Works*, I, p. 204 (IV, p. 96), and *Novum Organum*, II, Aph. 10, *Works*, I, p. 236 (IV, p. 127): 'Yet Natural and Experimental History is so various and diffuse, that it confounds and distracts the intellect, unless it be ranged and presented to view in a suitable order. Therefore we must form *Tables and Arrangements of Instances*, in such a method and order that the intellect may be able to deal with them.'

48 And see *Novum Organum*, I, Aph. 103, *Works*, I, p. 210 (IV, p. 96). Cf. 'Experientia literata', in L. Jardine, *Francis Bacon: Discovery and the Art of Discourse* (1974), pp. 143–9.

49 And see *Novum Organum*, I, Aphorismi 104, 106; *Works*, I, pp. 204–5, 206 (IV, pp. 97, 98).

50 'Distributio operis', *Instauratio magna*, *Works*, I, p. 138 (IV, p. 26).

51 *Novum Organum*, I, Aph. 105, *Works*, I, p. 205 (IV, p. 97); and see *Novum Organum*, II, Aph. 15, *Works*, I, pp. 256–7 (IV, p. 145):

> To God (the giver and artificer of Forms) and perhaps to angels and higher intelligences, it belongs to have an affirmative knowledge of Forms immediately, and from the first contemplation. But certainly this is beyond man; to whom it is granted merely to proceed at first by Negatives, and at last to end in Affirmatives, after every sort of exclusion.

52 *Novum Organum*, II, Aphorismi 15, 16, *Works*, I, pp. 256–7 (IV, pp. 145–6); and *Novum Organum*, II, Aph. 3, *Works*, I, pp. 228–9 (IV, p. 120).
53 *Novum Organum*, II, Aph. 18, *Works*, I, p. 259 (IV, p. 148).
54 *Novum Organum*, II, Aph. 21, *Works*, I, p. 268 (IV, p. 155).
55 *Novum Organum*, II, Aph. 52, *Works*, I, pp. 363–4 (IV, pp. 246–7).
56 *Novum Organum*, II, Aph. 32, *Works*, I, p. 286 (IV, p. 173).
57 *Parasceve*, Aph. 5, *Works*, I, p. 399 (IV, p. 258).
58 *Novum Organum*, II, Aph. 52, *Works*, I, p. 364 (IV, p. 247).
59 *Novum Organum*, II, Aph. 20, *Works*, I, pp. 261–2 (IV, p. 150).
60 *Ibid.*, *Works*, I, pp. 262–5 (IV, pp. 151–4).
61 *Ibid.*, *Works*, I, p. 266 (IV, pp. 154–5).
62 How this decision was arrived at is nowhere discussed.
63 *Novum Organum*, II, Aph. 21, *Works*, I, p. 268 (IV, p. 155).
64 *Parasceve*, *Works*, I, p. 393 (IV, pp. 251–2).
65 *Novum Organum*, II, Aph. 17, *Works*, I, pp. 257–8 (IV, p. 146).
66 *Novum Organum*, II, Aph. 2, *Works*, I, p. 228 (IV, p. 120). 'Whosoever knows the Forms, comprehends the unity of nature in the most dissimilar substances', *Novum Organum*, II, Aph. 3, *Works*, I, p. 229 (IV, p. 120).
67 *Ibid.*
68 *Advancement of Learning*, *Works*, III, p. 379.
69 *Parasceve*, *Works*, I, p. 403 (IV, p. 263).
70 *Ibid.*, p. 402 (IV, p. 261).
71 *Ibid.*
72 'Distributio operis', *Instauratio magna*, *Works*, I, pp. 138–9 (IV, p. 26).
73 Many references, e.g., *Parasceve*, Aph. 1, *Works*, I, p. 395 (IV, p. 253); *Advancement of Learning*, *Works*, III, p. 333.
74 *Parasceve*, Aph. 5, *Works*, I, pp. 398–9 (IV, p. 257).
75 *Parasceve*, Aph. 8, *Works*, I, p. 401 (IV, pp. 259–60).
76 *Oxford English Dictionary*, *s.v.* 'compearance'.
77 *Advancement of Learning*, *Works*, III, p. 363.
78 Cicero, *Topica*, X, 41–3; *De inventione*, I, 31–44; Quintilian, *Institutio oratoria*, V, x, 73.
79 *De augmentis scientiarum*, VIII, Aph. 76, *Works*, I, p. 821 (V, p. 104).
80 *Advancement of Learning*, *Works*, III, p. 330.
81 *Ibid.*, p. 353.

82 *Ibid.*, pp. 347–9.
83 'Valerius Terminus', *Works*, III, pp. 230–1.

Conclusion

1 Reported by John Aubrey, *Brief Lives*, ed. A. Clark (1898), I, p. 299.
2 See J. Martin, 'Natural Philosophy and its Public Concerns', in S. Pumfrey, P. Rossi and M. Slawinski (eds.), *Science, Culture and Popular Belief in Renaissance Europe* (1991), pp. 100–18.
3 *Doctor Faustus*, I, i, lines 53–5.
4 *The First Anniversarie*, lines 213–18.
5 See C. Webster, *The Great Instauration: Science, Medicine and Reform, 1626–1660* (1975).
6 See M. Hunter, *Science and Society in Restoration England* (1981).
7 Bacon to King James in 1612; *Letters*, IV, p. 280.
8 M. Hunter, *Science and Society*, pp. 8–31.
9 S. Shapin and S. Schaffer, *Leviathan and the Air-Pump: Hobbes, Boyle and the Experimental Life* (1985).
10 Christiaan Huygens to Colbert in 1666, in Huygens, *Œuvres completes*, VI (1895), pp. 95–6; in the margins of each paragraph of this letter, Colbert had marked *bon*. More generally, see R. Hahn, *The Anatomy of a Scientific Institution: The Paris Academy of Sciences, 1666–1803* (1971), Chapter 1; and A. Stroup, *A Company of Scientists: Botany, Patronage, and Community at the Seventeenth-Century Parisian Royal Academy of Sciences* (1990).

Bibliography

Unprinted sources

British Library: Lansdowne MS 101; Lansdowne MS 1121, fos. 112–20v; Stowe MS 424, fos. 145–50

Cambridge University Library: C.U.L. MS Ff. 4.16

Public Record Office:

State Papers, Domestic: Elizabeth I [SP. 12]; James I [SP. 14]

Law reports

The All England Reports. London 1930–

The English Reports. 176 vols. London 1900–30

Anderson [vol. 123]

Coke [vols. 76, 77]

Moore (K.B.) [vol. 72]

Plowden [vol. 75]

Popham [vol. 79]

Yelverton [vol. 80]

Haywarde, John *Les Reportes del Cases in Camera Stellata*, ed. W.P. Baildon. London 1894

Editions of works by Francis Bacon

The Works of Francis Bacon, eds. J. Spedding, R.L. Ellis, D.D. Heath. 7 vols. London 1857–61

The Letters and the Life of Francis Bacon, ed. J. Spedding. 7 vols. London 1861–74

A Conference of Pleasure, ed. J. Spedding. London 1870

'An Advertisement Touching Private Censure', in *Alnwick MS*, ed. F. Burgoyne. London 1904

The Essayes or Counsels, Civill and Morall, ed. M. Kiernan. Oxford 1985

The Essays, ed. J. Pitcher. Harmondsworth 1985

Bibliography

Commentaries on Bacon cited

Anderson, F.M. *The Philosophy of Francis Bacon*. Chicago 1948
Francis Bacon: His Career and His Thought. Los Angeles 1962
Briggs, J.C. *Francis Bacon and the Rhetoric of Nature*. Cambridge, Mass. 1989
Church, R.W. *Bacon*. London 1884
Crowther, J.G. *Francis Bacon: The First Statesman of Science*. London 1960
Epstein, J. *Francis Bacon: A Political Biography*. Athens, Ohio 1977
Farrington, B. *Francis Bacon: Philosopher of Industrial Science*. New York 1949
The Philosophy of Francis Bacon: An Essay on its Development from 1603 to 1609 with new translations of fundamental texts. Liverpool 1964
Gardiner, S.R. 'Francis Bacon', in *Dictionary of National Biography*, ed. L. Stephens. London 1885
Hesse, M. 'Francis Bacon', in *Dictionary of Scientific Biography*, ed. C.C. Gillispie. New York 1970
Jardine, L.A. *Francis Bacon: Discovery and the Art of Discourse*. Cambridge 1974
Knights, L.C. *Explorations*. London 1946
Kocher, P.H. 'Francis Bacon on the Science of Jurisprudence', *Journal of the History of Ideas*, 18 (1957), pp. 3–26
Macaulay, T.B. 'Lord Bacon', *Edinburgh Review*. July 1837
Marwil, J. *The Trials of Counsel: Francis Bacon in 1621*. Detroit 1976
Neustadt, M. 'The Making of the Instauration'. Unpublished Ph.D. thesis, The Johns Hopkins University 1987
Perez-Ramos, A. *Francis Bacon's Idea of Science and the Maker's Knowledge Tradition*. Oxford 1988
Rabb, T.K. 'Francis Bacon and the Reform of Society', in *Action and Conviction in Early Modern Europe*, eds. T.K. Rabb and J. Seigel. Princeton, N.J. 1969
Rees, G. and C. Upton *Francis Bacon's Natural Philosophy: A New Source*. Chalfont St Giles 1984
Rossi, P. *Francis Bacon: From Magic to Science*. trans. S. Rabinovitch. Chicago 1968
Shapiro, B.J. 'Sir Francis Bacon and the Mid-Seventeenth Century Movement for Law Reform', *American Journal of Legal History*, 24 (1980), pp. 331–62
Simonds, R.T. 'Bacon's Legal Learning: Its Influence on His Philosophical Ideas', pp. 493–501 in *Acta Conventus Neo-Latini Sanctandreani: Proceedings of the Fifth International Congress of Neo-Latin Studies*, ed. I.D. McFarlane. New York 1986
Urbach, P. *Francis Bacon's Philosophy of Science: An Account and a Reappraisal*. La Salle, Ill. 1987
Vickers, B. *Francis Bacon and Renaissance Prose*. Cambridge 1968
Vickers, B. (ed.) *Essential Articles for Studying Francis Bacon*. Hamden, Conn. 1968
Wallace, K. *Francis Bacon on Communication and Rhetoric*. Chapel Hill 1943

Whitney, C. *Francis Bacon and Modernity*. New Haven 1986
Williamson, G. *The Senecan Amble: A Study in Prose Form from Bacon to Collier*. London 1951
Wood, P. 'Francis Bacon and the "Experimentall Philosophy": A Study in Seventeenth-century Methodology'. Unpublished M.Phil. thesis, University of London 1978
Zetterburg, P.J. 'Echoes of Nature in Solomon's House', *Journal of the History of Ideas*, 43 (1982), pp. 179–93

Early modern texts and editions

(Original capitalisation – or lack of it – is retained in texts printed in early modern England, and *Short-Title Catalogue* numbers are included where appropriate.)

[*Anon.*] *A Dialogue, concerning the strife of our churche: wherein are aunswered diuers accusations, wherewith the godly are falsly charged*. London 1584 (*S.T.C.* 6801)
[*Anon.*] *A petition directed to her most excellent maiestie*. Middleburg 1591 (*S.T.C.* 1521)
Acts of the Privy Council of England, ed. J.R. Dasent. London 1880–1907
Aubrey, John *Brief Lives*, ed A. Clark. 2 vols. Oxford 1898
B[?]. R. *The difference between the Ancient Physicke ... and the latter Physicke*. London 1585 (*S.T.C.* 1064)
Bancroft, Richard *A Sermon preached at Paules Cross the 9. of Februarie, 1588*. London 1589 (*S.T.C.* 1346)
Daungerous positions and proceedings, published and practised within this iland of Brytaine. London 1593 (*S.T.C.* 1344)
Banister, John *A Compendious Chyrurgerie*. London 1585 (*S.T.C.* 25185)
Bowyer, Robert *The Parliamentary Diary of Robert Bowyer*, ed. D.H. Willson. Minneapolis 1931
Brooke, Sir Robert *La Graunde Abridgement*. London 1573 (*S.T.C.* 3827)
Budé, Guillaume *De L'institution du prince*. Paris 1547
Calendar of the Salisbury Manuscripts. Historical Manuscripts Commission. 24 vols. London 1883–1976
Calendar of State Papers, Foreign, Elizabeth, ed. W.B. Turnbull et al. 23 vols. London 1863–1950
Camden, William *Annals, or, the historie*, trans. R. Norton. London 1635 (*S.T.C.* 4501)
Castiglione, Baldesar *The Courtyer*, trans. Sir Thomas Hoby. London 1561 (*S.T.C.* 4778)
Certain Sermons, or Homilies. London 1587 (*S.T.C.* 13657)
Chaloner, Sir Thomas *A Shorte Discourse of the most rare and excellent vertue of Nitre*. London 1584 (*S.T.C.* 4940)
Coke, Sir Edward *The First Part of the Institutes of the Laws of England*. London 1628 (*S.T.C.* 15784)
The Fourth Part of the Institutes of the Laws of England. London 1648 (*S.T.C.* Wing. C4961)

Bibliography

Commons Debates, 1621, eds. W. Notestein, F.H. Relf and H. Simpson. Volume VII. New Haven 1935

Cooper, Thomas *An Admonition to the People of England*, ed. W.Goode. Cambridge 1850

Cowell, John *The Interpreter*. Cambridge 1607 (*S.T.C.* 5900)

Dallington, Sir Richard *Hypnerotomachia*. London 1592 (*S.T.C.* 5577)

De Guevara, Antonio *The Diall of Princes*, trans. Thomas North. London 1557 (*S.T.C.* 12427)

D'Ewes, Sir Simonds *The Journals of all the Parliaments during the Reign of Queen Elizabeth, both of the House of Lords and House of Commons*. London 1682 (*S.T.C.* Wing. D1250)

Digges, Thomas *A Geometrical practise, named Pantometria*. London 1571 (*S.T.C.* 6858)

Donne, John *The Complete Poetry*, ed. J.T. Shawcross. New York 1967

Du Bartas, Sallust *Works*, ed. U.T. Holmes et al. 3 vols. Chapel Hill 1938

Edward VI *The Chronicle and Political Papers of King Edward VI*, ed. W.K. Jordan. Ithaca, 1966

Elyot, Sir Thomas *Dictionary*, ed. Thomas Cooper. London 1548 (*S.T.C.* 7661)
The Booke Named The Governour. London 1531 (*S.T.C.* 7635)

Erasmus, Desiderius *Institutio Principis Christiani*. Basel 1514

Fitzherbert, Sir Anthony *Le Graunde Abridgement*. London 1565 (*S.T.C.* 10956)

Fraunce, Abraham *Lawiers Logike, exemplifying the praecepts of logike by the practise of the common law*. London 1588 (*S.T.C.* 11343)

Fulbeck, William *Direction, or Preparative to the study of the lawe*. London 1600 (*S.T.C.* 11410)

Gesta Grayorum, ed. D.S. Bland. Liverpool 1968

The Register of Admissions to Gray's Inn, 1521–1589, ed. J. Foster. London 1889

The Pension Book of Gray's Inn, 1569–1800, ed. R.J. Fletcher. 2 vols. London 1901–10

Hake, Edward *Epieikeia, a Dialogue on Equity* (*c.* 1595), ed. D.E.C. Yale. New Haven 1953

Harvey, Richard *A theologicall discourse of the Lamb of God and his enemies*. London 1589 (*S.T.C.* 12915)

Hastings, Francis *The Letters of Francis Hastings, 1574–1609*, ed. C. Cross. Frome 1969

Hobbes, Thomas *De Cive. The English Version*, ed. H. Warrender. Oxford 1983

Hooker, Richard *Of the lawes of ecclesiasticall politie*. London 1593 (*S.T.C.* 13712)

Hudson, William 'A Treatise on the Court of Star Chamber', *Collectanea Juridica*, ed. F. Hargrave. Volume II. London 1792

Huygens, Christiaan *Œuvres completes*. 22 vols. in 23. La Haye 1888–1950

James I *The Political Works of James I*, ed. C.H. McIlwain. Cambridge, Mass. 1918

Kellwaye, Simon *A Defensative Against the Plague*. London 1593 (*S.T.C.* 14917)

Littleton, Sir Thomas *Tenures*, ed. E. Wamborough. Washington 1903

Marlowe, Christopher *Complete Plays and Poems*, eds. E.D. Pendry and J.C. Maxwell. London 1976
Marsiglio (Marsilius) of Padua *The Defence of Peace*, trans. William Marshall. London 1535 (*S.T.C.* 17817)
The Defender of the Peace, ed. and trans. A. Gewirth. 2 vols. New York 1951, 1956
Marten, Anthony *A second sound, or warning of the trumpet unto judgement*. London 1589 (*S.T.C.* 17491)
A reconciliation of all the pastors and cleargy of this church of England. London 1590 (*S.T.C.* 17490)
Mascall, Leonard *The first booke of cattell*. London 1587 (*S.T.C.* 17580)
A booke of fishing with hooke & Line. Another of sundrie engines and trappes. London 1590 (*S.T.C.* 17572)
Newton, Thomas *The olde mans dietarie*. London 1586 (*S.T.C.* 18513)
Northamptonshire Lieutenancy Papers, 1580–1614, eds. J. Goring and J. Wake. Gateshead 1975
Partridge, John *The widowes treasure, plentifully furnished with secretes in physicke*. London 1585 (*S.T.C.* 19433.3)
Patrizi, Francesco *De regno et regis institutione*. Prato 1531
Paulet, Sir Amias *Copy-Book of Letters*, ed. O. Ogle. London 1866
Payne, Robert *The vale mans table. Herein is taught ... howe to draine moores*. London 1583 (*S.T.C.* 19491.5)
Peele, George *An Eglogue. Gratulatorie*. London 1589 (*S.T.C.* 19534)
Plat, Sir Hugh *A briefe Apology of certaine new inventions*. London 1593 (*S.T.C.* 19977)
Diverse new sorts of soyle. London 1594 (*S.T.C.* 19989)
The Jewell House of Art and Nature. London 1594 (*S.T.C.* 19991)
A discoverie of certaine English wants. London 1595 (*S.T.C.* 19988)
Plowden, Edmund *Les Commentaries*. 2 Parts. London 1571, 1579 (*S.T.C.* 20040, 20046.3)
Proceedings in the Parliaments of Elizabeth I. Volume I: 1559–1581, ed. T.E. Hartley. Leicester 1981
St Germain, Christopher. *A Dialogue in Englisshe, bytweene a doctour of dyvynyte, and a student in the lawes of Englande*, eds. T.F.T. Plucknett and J.L. Barton. London 1974
The Second Part of a Register, ed. A. Peel. 2 vols. Cambridge 1915
[*Sidney Papers*] *Letters and Memorials of State ... by Sir Henry Sidney ... Sir Philip Sidney and ... Sir Robert Sidney*, ed. Arthur Collins. 2 vols. London 1746
Sidney, Sir Philip *An Apology for Poetry, or, The Defense of Poesy*, ed. G. Shepherd. Manchester 1967
Smith, Sir Thomas *A Discourse of the Commonweal of This Realm of England*, ed. M. Dewar. Charlottesville 1969
De Republica Anglorum, ed. M. Dewar. Cambridge 1982
Statutes of the Realm, eds. A. Luders, T.E. Tomlins, J. France et al. 12 vols. London 1810–28

Stow, John *A survay of London*. London 1598 (*S.T.C.* 23341)

Stuart Royal Proclamations: I The Proclamations of King James I, 1603–1625, eds. P.L. Hughes and J.F. Larkin. Oxford 1973

Taverner, John *Certaine experiments concerning fish and fruite*. London 1600 (*S.T.C.* 23708)

Theloall, Simon *Le digest des briefes originals*. London 1579 (*S.T.C.* 23934)

Townshend, Hayward *Historical Collections: An exact account of the last four parliaments of Elizabeth*. London 1680 (*S.T.C.* Wing. T1991)

Tudor Royal Proclamations, eds. P.L. Hughes and J.F. Larkin. 3 vols. New Haven 1964, 1969

Whitgift, John *Works*, ed. J. Ayre. 3 vols. Cambridge 1851

Other books and articles cited

Abbott, L.W. *Law Reporting in England, 1485–1585*. London 1973

Allen, C.K. *Legal Duties and Other Essays in Jurisprudence*. Oxford 1931

Law in the Making. 7th edition. Oxford 1964

Allen, J.W. *English Political Thought, 1603–1660*. Volume I. London 1938

Andrews, K.R. *Trade, Plunder and Settlement: Maritime Enterprise and the Genesis of the British Empire , 1480–1630*. Cambridge 1984

Aristotle, *Topica*, trans. E.S. Forster. Cambridge, Mass. 1960

Arnold, M.S., T.A. Green, S.A. Scully and S.D. White (eds.) *On the Laws and Customs of England: Essays in Honor of Samuel E. Thorne*. Chapel Hill 1981

Baker, J.H. 'New Light on Slade's Case', *Cambridge Law Journal*, 29 (1971), pp. 51–67, 213–36

English Legal Manuscripts. 2 vols. London 1975, 1978

An Introduction to English Legal History, 2nd edition. London 1979

The Order of Serjeants at Law. London 1984

The Legal Profession and the Common Law: Historical Essays. London 1986

Baker, J.H. (ed.) *The Reports of Sir John Spelman*. 2 vols. London 1978

Barnes, T.G. 'Due Process and Slow Process in the late Elizabethan and early Stuart Star Chamber', *American Journal of Legal History*, 6 (1962), pp. 221–49, 315–46

Bauckham, R. *Tudor Apocalypse: Sixteenth Century Apocalypticism, Millenarianism and the English Reformation*. Appleford 1978

Beal, P. *Index of English Literary Manuscripts. Volume 1, 1450–1625, Part 1: 'Andrewes–Donne'*. London 1980

Beier, A.L. *Masterless Men: The Vagrancy Problem in England, 1540–1640*. London 1985

Bell, H.E. *An Introduction to the History and Records of the Court of Wards and Liveries*. Cambridge 1953

Bellamy, J.G. *The Tudor Law of Treason: An Introduction*. London 1979

Criminal Law and Society in Late Medieval and Tudor England. Gloucester 1984

Bibliography

Berman, H.J. *Law and Revolution: The Formation of the Western Legal Tradition*. Cambridge, Mass. 1983

Bindoff, S.T. 'The Making of the Statute of Artificers', in *Elizabethan Government and Society*, eds. S.T. Bindoff, J. Hurstfield and C.H. Williams. London 1961

Bindoff, S.T. (ed.) *The House of Commons, 1509–1558*. 3 vols. London 1982

Bowen, C.D. *The Lion and the Throne: The Life and Times of Sir Edward Coke, 1552–1634*. London 1957

Brockliss, L.W.B. *French Higher Education in the Seventeenth and Eighteenth Centuries: A Cultural History*. Oxford 1987

Brooks, E. St John *Christopher Hatton*. London 1946

Burgess, P.G. 'Custom, Reason and the Common Law: English Jurisprudence, 1600–1650'. Unpublished Ph.D. thesis. Cambridge University 1988

Bush, D. *English Literature in the Earlier Seventeenth Century*. Oxford 1962

Butters, H.C. *Governors and Government in Early Sixteenth-Century Florence, 1502–1519*. Oxford 1985

Caspari, F. *Humanism and the Social Order*. New York 1954

Challis, C.E. *The Tudor Coinage*. Manchester 1978

Christianson, P. 'The Causes of the English Revolution: A Reappraisal', *Journal of British Studies*, 15 (1976), pp. 40–75

Cicero *De inventione* and *Topica*, ed. and trans. H.M. Hubbell. Cambridge, Mass. 1949

Clark, P. (ed.) *The General Crisis of the 1590s*. London 1985

Cockburn, J.S. *A History of the English Assizes, 1558–1714*. Cambridge 1971

Collinson, P. *The Elizabethan Puritan Movement*. London 1967

'Sir Nicholas Bacon and the Elizabethan Via Media', *Historical Journal*, 23 (1980), pp. 255–73

The Religion of Protestants: The Church in English Society, 1559–1625. Oxford 1982

Godly People: Essays on English Protestantism and Puritanism. London 1983

Cross, R. *Precedent in English Law*. Oxford 1977

Davies, M.G. *The Enforcement of English Apprenticeship: A Study in Applied Mercantilism, 1563–1642*. Cambridge, Mass. 1956

Dawley, P.M. *John Whitgift and the Reformation*. London 1955

Dewar, M. *Sir Thomas Smith: A Tudor Intellectual in Office*. London 1964

Dias, R.W.M. 'The Empirical Approach to the Teaching of Jurisprudence', *Law Quarterly Review*, 72 (1956), pp. 557–74

Dickens, A.G. *The English Reformation*. London 1967

Elton, G.R. *The Tudor Revolution in Government: Administrative Changes in the Reign of Henry VIII*. Cambridge 1953

Policy and Police: The Enforcement of the Reformation in the Age of Thomas Cromwell. Cambridge 1972

Reform and Renewal: Thomas Cromwell and the Commonweal. Cambridge 1973

Studies in Tudor and Stuart Politics and Government: Papers and Reviews, 1946–1972. 2 vols. Cambridge 1974
The Tudor Constitution: Documents and Commentary. 2nd edition. Cambridge 1982
Studies in Tudor and Stuart Politics and Government: Papers and Reviews, 1973–1981. Cambridge 1983
The Parliament of England, 1559–1581. Cambridge 1986
Elton, G.R. (ed.) *The New Cambridge Modern History*, Volume II: *The Reformation*. Cambridge 1958
Essays in History and Political Theory in Honor of Charles Howard McIlwain. Cambridge, Mass. 1936
Ferguson, A.B. *The Articulate Citizen and the English Renaissance*. Durham, N.C. 1965
Finlay, R. *Politics in Renaissance Venice*. London 1980
Gardiner, S.R. *The History of England from the Accession of James I to the Outbreak of the Civil War, 1603–1642*. 10 vols. London 1884
Gilbert, F. 'Bernardo Rucellai and the *Orti Oricellari*: A Study of the Origin of Modern Political Thought', *Journal of the Warburg and Courtauld Institutes*, 12 (1949), pp. 101–31
Giuliani, A. 'The Influence of Rhetoric on the Law of Evidence and Pleading', *Juridical Review*, 7 (1962), pp. 216–52
Gleason, J.H. *The Justices of the Peace in England, 1558–1640*. Oxford 1969
Gould, J.D. *The Great Debasement: Currency and the Economy in Mid-Tudor England*. Oxford 1970
Grafton, A. and L.A. Jardine *From Humanism to the Humanities: Education and the Liberal Arts in Fifteenth- and Sixteenth-Century Europe*. Cambridge, Mass. 1986
Green, T.A. *Verdict according to Conscience: Perspectives on the English Criminal Trial Jury, 1200–1800*. Chicago 1986
Guest, A.G. (ed.) *Oxford Essays in Jurisprudence*. Oxford 1961
Guth, D.J. and J.W. McKenna (eds.) *Tudor Rule and Revolution: Essays for G.R. Elton from his American Friends*. Cambridge 1982
Guy J.A. and H.G. Beale (eds.) *Law and Social Change in British History*. London 1984
Hahn, R. *The Anatomy of a Scientific Institution: The Paris Academy of Sciences, 1666–1803*. Berkeley 1971
Haigh, C. (ed.) *The Reign of Elizabeth*. London 1984
Hamilton, A.C. *Sir Philip Sidney: A Study of his Life and Works*. Cambridge 1977
Hasler, P. W. (ed.) *The House of Commons, 1558–1603*. 3 vols. London 1981
Heath, J. *Torture and English Law: An Administrative and Legal History from the Plantagenets to the Stuarts*. Westport, Conn. 1982
Heinze, R.W. *The Proclamations of the Tudor Kings*. Cambridge 1976
Herrup, C.B. *The Common Peace: Participation and the Criminal Law in Seventeenth-Century England*. Cambridge 1987
Hill, J.E.C. *Society and Puritanism in Pre-Revolutionary England*. London 1964

Intellectual Origins of the English Revolution. Oxford 1965
Holdsworth, W.S. *A History of English Law*. 16 vols. London 1922–66
Howell, R., jr. 'The Sidney Circle and the Protestant Cause in Elizabethan Foreign Policy', *Renaissance and Modern Studies*, 19 (1975), pp. 31–46
Hulme, E.W. 'The History of the Patent System under the Prerogative and at Common Law', *Law Quarterly Review*, 12 (1896), pp. 141–54; 16 (1900), pp. 44–56
Hunter, M. *Science and Society in Restoration England*. Cambridge 1981
The Royal Society and its Fellows, 1660–1700: The Morphology of an Early Scientific Institution. Chalfont St Giles 1982
Ives, E.W. 'The Purpose and Making of the Later Year Books', *Law Quarterly Review*, 89 (1973), pp. 64–88
The Common Lawyers of Pre-Reformation England: Thomas Kebell: A Case Study. Cambridge 1983
Jones, W.J. *The Elizabethan Court of Chancery*. Oxford 1967
Jones, W.R.D. *The Tudor Commonwealth, 1529–1559*. London 1970
Joy, L.S. *Gassendi the Atomist: Advocate of History in an Age of Science*. Cambridge 1988
Judson, M. *The Crisis of the Constitution*. Brunswick, N.J. 1949
Kantorowicz, E.H. *The King's Two Bodies: A Study in Mediaeval Political Theology*. Princeton, N.J. 1957
Kargon, R. *Atomism in England: Hariot to Newton*. Oxford 1966
Kelley, D.R. 'History, English Law and the Renaissance', *Past and Present*, 65 (1974), pp. 24–51
Kent, J. 'Attitudes of Members of the House of Commons to the Regulation of "Personal Conduct" in Late Elizabethan and Early Stuart England', *Bulletin of the Institute of Historical Research*, 46 (1973), pp. 43–71
King, B.E. 'The Concept of a Lawyer's Jurisprudence', *Cambridge Law Journal*, 11 (1952–3), pp. 230–9 and pp. 404–20
Kishlansky, M. 'The Emergence of Adversary Politics in the Long Parliament', *Journal of Modern History*, 49 (1977), pp. 617–40
Klibansky, R., E. Panofsky and F. Saxl *Saturn and Melancholy: Studies in the History of Natural Philosophy, Religion and Art*. London 1964
Knafla, L.A. *Law and Politics in Jacobean England: The Tracts of Lord Chancellor Ellesmere*. Cambridge 1977
'The Influence of Continental Humanists and Jurists on English Common Law in the Renaissance', pp. 60–71 in *Acta Conventus Neo-Latini Bononiensis: Proceedings of the Fourth International Congress of Neo-Latin Studies*, ed. R.J. Shoeck. New York 1985
Knox, S.J. *Walter Travers: Paragon of Elizabethan Puritanism*. London 1962
Kristeller, P.O. *Iter Italicum: A Finding List of Uncatalogued or Incompletely Catalogued Humanistic Manuscripts of the Renaissance in Italian and other Libraries*. Vol. II. London 1967
Langbein, J.H. *Prosecuting Crime in the Renaissance: England, Germany, France*. Cambridge, Mass. 1974

Torture and the Law of Proof: Europe and England in the Ancien Regime. Chicago 1977
Levack, B.P. *The Civil Lawyers in England, 1603–1642: A Political Study.* Oxford 1973
Levi, E.H. *An Introduction to Legal Reasoning.* Chicago 1949
Lewis, J.U. 'Coke's Theory of Artificial Reason', *Law Quarterly Review*, 84 (1968), pp. 330–42
Lewis, T.E. 'The History of Precedent', *Law Quarterly Review*, 46, 47, 48 (1930–2)
Lloyd, D. 'Reason and Logic in the Common Law', *Law Quarterly Review*, 64 (1948), pp. 468–84
Loach, J. and R. Tittler (eds.) *The Mid-Tudor Polity, c. 1540–1560.* London 1980
Lossee, J. *A Historical Introduction to the Philosophy of Science.* Oxford 1980
MacCaffrey, W.T. *The Shaping of the Elizabethan Regime.* Princeton, N.J. 1968
Queen Elizabeth and the Making of Policy, 1572–1588. Princeton, N.J. 1981
MacLeod, C. *Inventing the Industrial Revolution: The English Patent System, 1660–1800.* Cambridge 1988
Maclure, M. *The Paul's Cross Sermons, 1534–1642.* Toronto 1958
Maitland, F.W. (ed.) *Year Book of 3 Edward II.* Vol. III. London 1905
Manuel, F.E. and F.P. Manuel *Utopian Thought in the Western World.* Oxford 1979
McConica, J.K. *English Humanists and Reformation Politics.* Oxford 1965
'Humanism and Aristotle in Tudor Oxford', *English Historical Review*, 94 (1979), pp. 291–317
McIlwain, C.H. *The High Court of Parliament and Its Supremacy.* New Haven 1910
Milsom, S.F.C. 'Law and Fact in Legal Development', *University of Toronto Law Journal*, 17 (1967), pp. 1–19
The Legal Framework of English Feudalism. Cambridge 1976
Historical Foundations of the Common Law. London 1981
Studies in the History of the Common Law. London 1986
Morgan J. *Godly Learning: Puritan Attitudes towards Reason, Learning and Education, 1560–1640.* Cambridge 1986
Mosse, G. *The Struggle for Sovereignty in England, from the Reign of Queen Elizabeth to the Petition of Right.* East Lansing, Mich. 1950
Neale, J.E. *Queen Elizabeth.* London 1934
The Elizabethan House of Commons. London 1949
Queen Elizabeth I and Her Parliaments. 2 vols. London 1953, 1957
Notestein, W. *The Winning of the Initiative by the House of Commons.* London 1924
Oldroyd, D. *The Arch of Knowledge: An Introductory Study of the History of Philosophy and Methodology of Science.* London 1986
Osborn, J.M. *Young Philip Sidney, 1572–1577.* New Haven 1972

Palliser, D.M. *The Age of Elizabeth: England under the Later Tudors, 1547–1603*. London 1983
Phillips, J.E. and D.C. Allen *Neo-Latin Poetry of the Sixteenth and Seventeenth Centuries*. Los Angeles 1965
Pierce, W. *Historical Introduction to the Martin Marprelate Tracts*. London 1908
Pindar *Odes*, ed. and trans. J.E. Sandys. Cambridge, Mass. 1937
Plato *Critias* and *Timaeus*, ed. and trans. R.G. Bury. Cambridge, Mass. 1929
Plucknett, T.F.T. *A Concise History of the Common Law*. London 1956
Pollock, F. and F.W. Maitland *The History of English Law Before the Time of Edward I*, ed. S.F.C. Milsom. 2 vols. Cambridge 1968
Porter, H.C. *Reformation and Reaction in Tudor Cambridge*. Cambridge 1958
Pound, R. 'The Maxims of Equity: I. Of Maxims Generally', *Harvard Law Review*, 34 (1921), pp. 806–36
Prest, W.R. *The Inns of Court under Elizabeth I and the Early Stuarts, 1590–1640*. London 1972
Pulman, M.B. *The Elizabethan Privy Council in the Fifteen-seventies*. Berkeley 1971
Pumfrey, S., P. Rossi and M. Slawinski (eds.) *Science, Culture and Popular Belief in Renaissance Europe*. Manchester 1991
Quintilian *Institutio oratoria*, ed. and trans. H.E. Butler. 4 vols. London 1920
Read, C. *Mr Secretary Walsingham and the Policy of Queen Elizabeth*. 3 vols. Oxford 1925
Mr Secretary Cecil and Queen Elizabeth. London 1955
Lord Burghley and Queen Elizabeth. London 1960
Read, C. (ed.) *William Lambarde and Local Government*. Ithaca, N.Y. 1962
Richardson, H.G. and G.O. Sayles *Law and Legislation from Aethelbert to Magna Carta*. Edinburgh 1966
Richardson, W.C. *Tudor Chamber Administration, 1485–1547*. Baton Rouge 1952
The History of the Court of Augmentations, 1536–1554. Baton Rouge 1962
A History of the Inns of Court. Baton Rouge (n.d. 1977?)
Rosenberg, E. *Leicester: Patron of Letters*. New York 1955
Rossi, P. *Philosophy, Technology and the Arts in the Early Modern Era*, trans. S. Attanasio. London 1970
Rubinstein, N. (ed.) *Florentine Studies*. London 1968
Russell, C. 'Parliamentary History in Perspective, 1604–1629', *History*, 61 (1976), pp. 1–27
Sanders, G. (ed.) *Orders of the High Court of Chancery*. London 1845
Schmitt, C.B. *Studies in Renaissance Philosophy and Science*. London 1981
Aristotle and the Renaissance. Cambridge, Mass. 1983
John Case and Aristotelianism in Renaissance England. Kingston, Ont. 1983
Schramm, P.E. *A History of the English Coronation*, trans. L.G. Wickham Legg. Oxford 1937
Scofield, C.L. *A Study of the Court of Star Chamber*. Chicago 1900

Shapin, S. and S. Schaffer *Leviathan and the Air-Pump: Hobbes, Boyle and the Experimental Life*. Princeton, N.J. 1985

Shapiro, B.J. 'Law Reform in 17th Century England', *American Journal of Legal History*, 19 (1975), pp. 280–312

Probability and Certainty in Seventeenth-Century England. Princeton, N.J. 1983

A Short Title Catalogue ... 1475–1640, 2nd edition. N. Pantzer. 2 vols. London 1976, 1986

A Short Title Catalogue ... 1640–1700. ed. D. Wing. 3 vols. New York 1945–51

Simpson, A. *The Wealth of the Gentry, 1540–1600*. Cambridge 1961

Simpson, A.W.B. *A History of the Land Law*. 2nd edition. Oxford 1986

Skinner, Q. *The Foundations of Modern Political Thought*. 2 vols. Cambridge 1978

Slack, P. 'Vagrants and Vagrancy in England, 1598–1664', *Economic History Review*, 2nd series, 27 (1974), pp. 360–79

'Books of Orders: The Making of English Social Policy, 1577–1631', *Transactions of the Royal Historical Society*, 5th series, 30 (1979), pp. 1–22

Smith, A.H. *County and Court: Government and Politics in Norfolk, 1558–1603*. Oxford 1974

Somerville, R. *History of the Duchy of Lancaster*. Vol. I. London 1953

Stein, P. *Regulae Iuris: From Juristic Rules to Legal Maxims*. Edinburgh 1966

Legal Institutions: The Development of Dispute Settlement. London 1984

Stone, J. *Legal System and Lawyers' Reasonings*. London 1964

Precedent and Law: Dynamics of Common Law Growth. London 1985

Stone, L. *An Elizabethan: Sir Horatio Palavicino*. Oxford 1956

Stroup, A. *A Company of Scientists: Botany, Patronage, and Community at the Seventeenth-Century Parisian Royal Academy of Sciences*. Berkeley 1990

Taylor, E.G.R. *The Mathematical Practitioners of Tudor and Stuart England*. Cambridge 1954

Thirsk, J. *Economic Policy and Projects: The Development of a Consumer Society in Early Modern England*. Oxford 1978

Thirsk, J. (ed.) *The Agrarian History of England and Wales* Volume IV: 1500–1640. Cambridge 1967

Thorne, S.E. (ed.) *Readings and Moots at the Inns of Court in the Fifteenth Century*. Vol. I. London 1974

Thorne, S.E. and J.H. Baker (eds.) *Readings and Moots at the Inns of Court in the Fifteenth Century*. Vol. II. London 1990

Tittler, R. *Nicholas Bacon: The Making of a Tudor Statesman*. Athens, Ohio 1976

Turner, R.V. *The English Judiciary in the Age of Glanvill and Bracton, c. 1176–1239*. Cambridge 1985

Usher, R.G. *The Rise and Fall of the High Commission*, 2nd edition by P. Tyler. Oxford 1968

Van Caenegem, R.C. *The Birth of the Common Law*. Cambridge 1973

Judges, Legislators and Professors: Chapters in European Legal History. Cambridge 1987
Van Dorsten, J.A. *Poets, Patrons, and Professors: Sir Philip Sidney, Daniel Rogers, and the Leiden Humanists.* Leiden 1962
Virgil *Aeneid*, ed. and trans. H.R. Fairclough. 2 vols. London 1929
Walter, J. '"A Rising of the People"? The Oxfordshire Rising of 1596', *Past and Present*, 107 (1985), pp. 90–143
Walter, J. and K. Wrightson 'Dearth and the Social Order in Early Modern England', *Past and Present*, 71 (1976), pp. 22–42
Webster, C. *The Great Instauration: Science, Medicine and Reform, 1626–1660.* London 1975
Webster, C. (ed.) *Health, Medicine and Mortality in the Sixteenth Century.* Cambridge 1979
Wernham, R.B. *After the Armada: Elizabethan England and the Struggle for Western Europe, 1588–1595.* Oxford 1984
White, S.D. *Sir Edward Coke and the 'Grievances of the Commonwealth'.* Chapel Hill 1979
Wilks, M. *The Problem of Sovereignty in the Later Middle Ages.* Cambridge 1963
Williams, F.B. (ed.) *An Index of Dedications and Commendatory Verses in English Books Before 1641.* London 1962
Williams, P. *The Council in the Marches of Wales under Elizabeth.* Cardiff 1958
Wrightson, K. 'The Puritan Reformation of Manners'. Unpublished Ph.D. thesis. Cambridge University 1973
English Society, 1580–1680. London 1982
Yates, F.A. *The French Academies of the Sixteenth Century.* London 1947
The Occult Philosophy in the Elizabethan Age. London 1979
Youngs, F.A. *The Proclamations of the Tudor Queens.* Cambridge 1976
Zeeveld, W.G. *Foundations of Tudor Polity.* Cambridge, Mass. 1948

Index

Index

Index

Index

For EU product safety concerns, contact us at Calle de José Abascal, 56–1°, 28003 Madrid, Spain or eugpsr@cambridge.org.

www.ingramcontent.com/pod-product-compliance
Lightning Source LLC
LaVergne TN
LVHW090320210926
848480LV00001B/306